A Practical Guide to Vehicle Refinishing

A Practical Guide to Vehicle Refinishing

Julian Woodstock

CRC Press
Taylor & Francis Group
Boca Raton London New York

CRC Press is an imprint of the
Taylor & Francis Group, an **informa** business

CRC Press
Taylor & Francis Group
6000 Broken Sound Parkway NW, Suite 300
Boca Raton, FL 33487-2742

International Standard Book Number-13: 978-1-138-48666-9 (Hardback)
International Standard Book Number-13: 978-1-138-48664-5 (Paperback)

Library of Congress Cataloging-in-Publication Data

Names: Woodstock, Julian, author.
Title: A practical guide to vehicle refinishing / Julian Woodstock.
Description: Boca Raton : Taylor & Francis, [2020] | "A CRC title, part of
the Taylor & Francis imprint, a member of the Taylor & Francis Group, the
academic division of T&F Informa plc." | Includes bibliographical
references and index.
Identifiers: LCCN 2019014198 | ISBN 9781138486669 (hardback : alk. paper) |
ISBN 9781138486645 (pbk. : alk. paper) | ISBN 9781351040549 (ebook)
Subjects: LCSH: Automobiles—Painting—Handbooks, manuals, etc.
Classification: LCC TL255.2 .W66 2020 | DDC 629.2/60288—dc23
LC record available at https://lccn.loc.gov/2019014198

Visit the Taylor & Francis Web site at
http://www.taylorandfrancis.com

and the CRC Press Web site at
http://www.crcpress.com

Contents

Acknowledgements

Thank you to my friend John Kyle and my wife Lindsay Woodstock for all their help and support.

Author

Julian Woodstock delivers courses at levels 1, 2 and 3 in vehicle refinishing, restoration work and body repair at Colchester Institute, Essex, England, and has co-written much of the body repair and refinishing material currently available on the The Institute of the Motor Industry's training programme. He has considerable experience also at the sharp end of crash repair.

The logical sequence to refinishing

This book is designed mainly to help level 2 and level 3 refinishing students understand the technical side of their new subject, but it can also be used as a reference guide for refinishers already in the industry (believe it or not even the author is still trying new techniques and trialing new technology in the refinishing world which means learning all the time). Restoration and custom enthusiasts should also find many interesting features. The refinishing industry provides a rewarding career for many people. The satisfaction to be gained from seeing a vehicle come in with accident damage, go through the repair process, and then leave as a finished product is something that is not often found in other parts of the motor industry. Another rewarding aspect of working on vehicle paint work is the fact that every job is different. These differences may be working on different makes of vehicle, different panels, applying different colours and finishes, and obviously the different damage that may be found on the panels.

THE REFINISHING PROCESS

Part of the reason the author enjoys his career in refinishing so much is because no two jobs are the same. However, generally speaking, every job is going to involve some sanding, some priming and applying top coats; therefore, this book has been set out in the order in which repairs would be undertaken. Note that most jobs will not require all of the stages listed through the chapters.

Chapter 1

Identifying substrates

One of the first tasks to consider when undertaking any form of paint repair is to identify the substrate. This is very important because the substrate may need to be protected by the paint and may require special undercoats to provide adhesion or corrosion protection. Some substrates can react with top coats, whereas others can be damaged by cleaning agents or undercoats.

So what is a substrate? Well, substrates can be broken down into two main categories: parent substrates and original or repainted panel coatings.

1.1 PARENT SUBSTRATES

These substrates are the material that the panel or component is made from. For example, the majority of motor vehicle body panels are made from low carbon (or mild) steel. However, many other materials may be used to manufacture body panels including plastics, aluminium, composites and alloys.

Low carbon steel will require an undercoat with some form of anticorrosion material, such as an acid etch primer. The easiest way to identify a steel component is to use a magnet. Only ferrous metals will attract a magnet. Of the ferrous metals available, only low carbon steel and its similar alloys are used in body construction. Stainless steel has been used very occasionally (the DMC DeLorean most famously), but it is usually nonmagnetic. If there is rust present on the panel, like the one in Figure 1.1 this will prove that the panel must be made from low carbon steel and is about the only visual clue to be relied upon.

If the panel concerned is metal but nonmagnetic, it is most likely to be aluminium. This also requires an etch primer, but for the reason that aluminium naturally builds a protective oxide layer very quickly. In reality, this oxide layer is not well attached at a molecular level, and although paint can 'stick' to it well, the oxide layer will peel from the metal easily causing adhesion problems. Therefore, bare aluminium should be thoroughly abraded and cleaned with a solvent panel wipe, then an etch primer applied immediately. A simple visual check for corrosion, which appears as a white powdery substance, as pictured in Figure 1.2 can help to identify aluminium.

If the panel or component is obviously made of plastic, it is most likely to be a thermoplastic. Thermoplastics are plastics that can be warmed and reshaped (a useful benefit if the panel is dented or deformed). They are most easily identified by a code embossed on an unseen part of the component such as PP+EPDM (polypropylene and

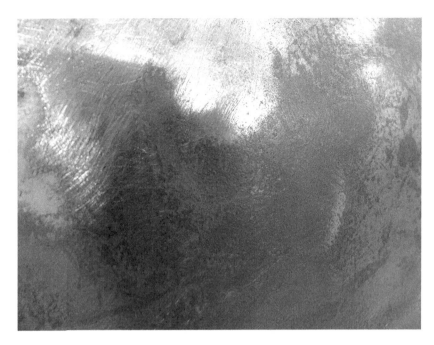

Figure 1.1 This panel is obviously very rusty, however the metal has not been made thinner by the corrosion. There is clean silver steel visible where the rust has been removed. This will then be treated with a rust converter before being primed and painted.

Figure 1.2 This alloy wheel has severe corrosion. The white powder is clearly visible on the bare metal areas.

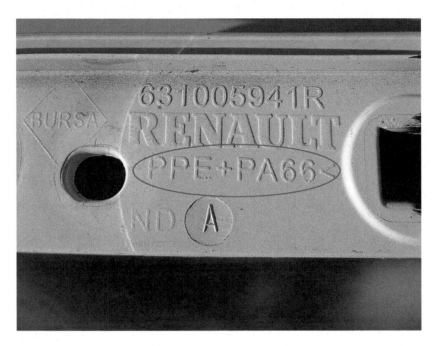

Figure 1.3 This plastic code was found on a Renault wing.

ethylene propylene diene monomer) as in Figure 1.3. If a code cannot be found, the information can usually be found in data manuals. Of course, if either of these is not forthcoming, an unseen part of the component can be warmed and attempts made to shape it, although this should be a last resort.

Plastic can be visually identified by sanding through the paint layers. Thermoplastics used in vehicle body components are usually black as in Figure 1.4, white, or grey; therefore, if such a substrate is found, it is likely to be a thermoplastic. These components can often be more flexible than metal components, so plasticiser (or elasticiser) should be added to the paint. More importantly though, the bare plastic should not be cleaned with a solvent panel wipe (water based should be used) as the solvent can be absorbed by the plastic and may show itself as Solvent Pop in the paint finish after baking. Finally, and possibly most importantly, bare thermoplastic requires the application of an adhesion promotor to ensure that any undercoats applied over the repairs remain stuck to the component.

Thermoset plastics and composites are not used as frequently as thermoplastics, and they can be identified in a number of different ways. As has already been discussed, thermoplastics can be warmed and reshaped. It should therefore be obvious that thermoset plastics cannot be reshaped in such a way; they are indeed often very rigid. Composites are most easily identified in a non-painted state; therefore, if it is possible to view the reverse side, this will help with the identification. Composites in body construction are usually made of strands such as glass fibre or, more rarely and exotically, carbon fibre. Glass-reinforced plastic is most often made from a randomly arranged matting of glass fibre as can be seen in Figure 1.5 whereas carbon fibre is structured as a woven material. Both are set in resin, which is hard when cured. As a result, they can be identified when sanding by a strong and distinct odour, although

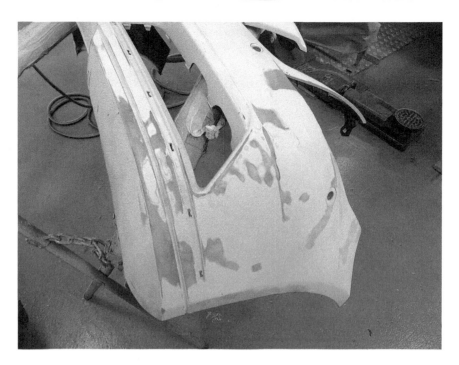

Figure 1.4 Here the white paint has been sanded back on this bumper to expose the black plastic. Note the pinky beige patch of body filler which has been applied to a damaged area.

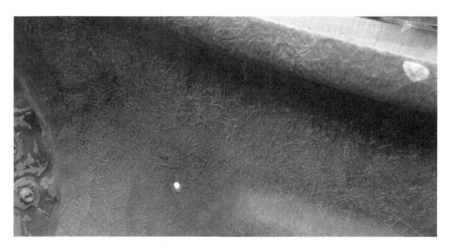

Figure 1.5 This panel was found in the front of a VW Beetle. The distinct random fibre on the reverse makes it easy to identify as glass reinforced plastic.

obviously a dust mask should be worn for all sanding operations. Care should be taken when sanding composites not to go through any gel coats that are present. Like thermoplastics, thermosets and composites should not be cleaned with a solvent panel wipe as it may evaporate through drying primer coats.

There are other parent substrates; however, the ones covered here are the most commonly found. Materials such as wood, tin and even fabric have been used in vehicle body construction over the years, but they are so rarely found we won't go over them here. The following are panel coatings that should be considered as substrates when being over coated.

1.2 ORIGINAL EQUIPMENT MANUFACTURER (OEM) PAINT MATERIAL

This is what the panel was originally painted in at the factory. The age of the vehicle should be considered, as there are a huge range of paint materials that could have been applied to the panels at the factory – different periods of production will indicate the most likely used material. These may include single-pack (1K) or two-pack (2K), high-bake enamels and ceramic clears. It is sometimes not entirely obvious whether the OEM paint is 1K, 2K, clear over base (COB), or direct gloss (DG); therefore, it is necessary to undertake a couple of tests to help identify these materials. It is also not always obvious whether the paint is OEM or not. You could perform a sanding test to see if there are other layers below the top one like in Figure 1.6, or defects such as clear peeling from poorly prepped paint as in Figure 1.7. However when presented with a nice clean example like Figure 1.8 it is pretty certain to be OEM.

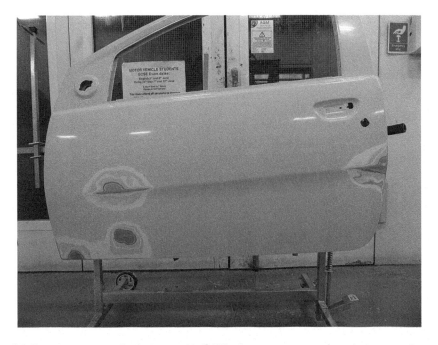

Figure 1.6 If you are unsure whether a panel is OEM paint or not, an un-obtrusive part can be sanded back to show the layers. This door is not OEM paint.

Figure 1.7 This panel is suffering from clear-coat peel where the OEM clear coat hadn't been properly sanded; this clearly marks it as previously painted.

Figure 1.8 This vehicle has clean OEM paint which is a good basis for repairs.

1.3 PREVIOUSLY REPAIRED PAINTWORK

If the panel has been previously refinished, it could have been repaired with any of the previously mentioned materials. As these materials react very differently to preparation and application of further coats, it is very important to distinguish what was used for the repair.

1.4 SOLVENT WIPE TEST

To decide whether a material is 1K or 2K, take a small amount of gun wash or thinners on a rag, and soak an area that is either going to be repaired or is out of sight. If the colour comes off on the rag, as it has in Figure 1.9 the panel is finished with 1K material. 1K paint will wrinkle on the panel if exposed to thinners for long enough. If there is no perceivable colour transfer to the rag, it is most likely that the panel is finished with 2K material. This isn't fool-proof, as heavily oxidised DG will still transfer to the cloth even if it is 2K; however, this will only be a small amount of paint oxide on the cloth, and the paint on the panel will not wrinkle up.

Figure 1.9 This wing has been painted sometime in the past. In order to ascertain if the paint is 1K or 2K, it is wiped with gun wash on a paper towel. The blue paint is clearly visible on the paper towel, which confirms that the paint is a 1K product.

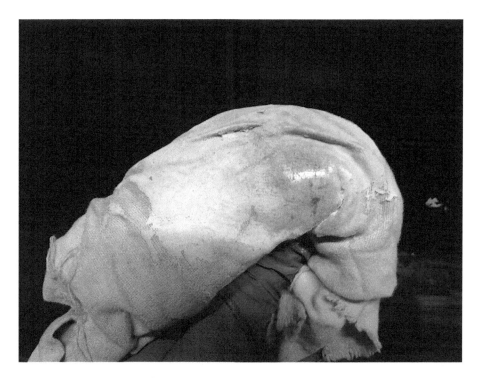

Figure 1.10 In order to ascertain whether a vehicle paint is DG or COB, it should be polished. If the colour of the vehicle is transferred to the polish on the rag, as shown in the picture, it confirms that the paint is DG.

1.5 POLISH AND SANDING TESTS

It may be relevant to differentiate between a DG and a COB finish. There are two procedures that may be used to help with discerning this. The first is to take a small amount of coarse polish on a mutton cloth and rub an area of the original finish. Again, if colour comes off onto the cloth, the finish is DG, and if there is no discernible colour as in Figure 1.10, it will be COB. The second type of test is quicker but a more destructive version of the first test and that is to use a fine-grade abrasive paper (P1000 for example) to flat the panel. Obviously, the dust (or sludge if wet flatting) will be the colour of the panel if it is finished with DG or white if it is COB.

1.6 CURRENT REPAIR FOUNDATION MATERIALS

These will be the undercoat materials that are used to repair and prepare the damaged panels to the correct standard that they need to be at for final top coats. These undercoats may be found in the foundation materials Section 1.3 but includes etch primer, non-sanding primer, high build primer (Figure 1.11) and filler. Replacement steel and aluminium panels are usually supplied pre-painted in an electrophoretic coating (e-coat) as in Figure 1.12. This has excellent adhesion to the metal

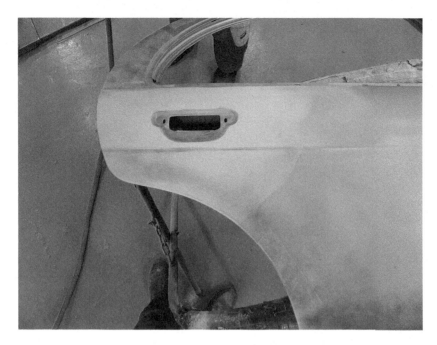

Figure 1.11 This patch of high-build primer is in the process of being sanded to a smooth consistent finish.

Figure 1.12 E-coat panels should be sanded and etch primed. This one is welded on so the back edges of the wing were painted silver before fitting.

Figure 1.13 Plastic components are often supplied in a very thin coat of primer. This should be sanded smooth prior to wet-on-wet primer being applied. This primer could although be sanded with a dual action sander however the primer on these plastic components the primer is usually very thin and easily removed therefore wet and dry is a good alternative, as hand sanding is more controllable.

component and a degree of corrosion resistance too. Therefore, when preparing e-coat, it is recommended as far as possible to not break through to steel. There are indeed many wet-on-wet primers currently available on the market, which simply require the e-coat to be cleaned (as long as there are no scratches in the e-coat). Other materials such as plastic are often supplied in a thin primer coat which can be lightly flatted with wet-and-dry, as in Figure 1.13, foam back pads or woven abrasive fleece.

Wherever possible, the substrates should be identified prior to the repair, and any materials used as undercoats should be considered throughout the undertaking. By considering and correctly identifying the substrates, problems with adhesion or reactions between materials can be avoided and therefore prevent costly and unnecessary rework.

Chapter 2

Identifying the damage

In most workshops, the damage will already be identified for you by the estimator, and the repair process will be roughly outlined in the job card. For the sake of this chapter though, it would be a good idea to discuss how to decide on the best techniques used to repair the damage such as that seen in Figure 2.1. Therefore, a few guidelines would be the following:

- Use a combination of visual and tactile inspection to check for dents, split paint, chips, corrosion, etc. Extra time spent at this point may well save far more time than will be spent reworking any missed damage down the line. It's good practice to do this while cleaning the panel either with soapy water or with panel wipe. The time taken to clean the panel(s) thoroughly can really help concentrate the technician's attention on checking for damage with the added bonus of then working on panels with no contamination present.

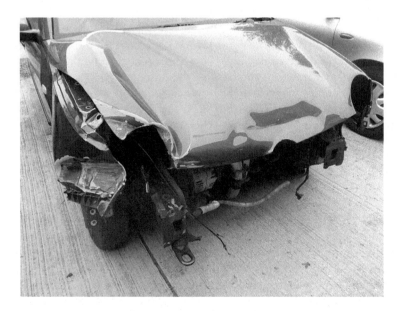

Figure 2.1 This Lupo has plenty of panels needing repair. The steel has a thin coat of galvanising prior to any primer being applied.

- Use the finest abrasive that will still deliver efficient preparation or removal. Coarse abrasives can leave deep scratches which will require further removal before over-coating with top coats or foundation coats; this in turn may cause surface deformation or promote reaction between existing undercoats and newly applied coatings. When preparing panels, the damage will often dictate what flatting techniques should be used. We'll discuss the use of wet flatting, dry flatting, machine sanding, hand sanding, etc. in Chapter 4.
- Perform a simple solvent wipe test and polish test as discussed in Chapter 1. This is very important when it comes to deciding how to repair the damage, as some materials are incompatible even when dry. As a general rule of thumb, single-pack materials are more unstable than two-pack materials. If there are any doubts, it is recommended to strip the paint material back to sound or parent substrate – unfortunately due to labour charges, etc., this is not often done. Instead, technicians are frequently instructed to isolate dubious previous paint with high-build primer before sanding it smooth. If we consider the various paint layers to be like a building, we should compare the first layers to be like the foundations. If these are unstable, the entire building (layers of paint) is likely to be compromised far sooner than if strong foundations are utilised.

2.1 TYPES OF DAMAGE

2.1.1 Stone chips

There are many ways to repair stone chips – they are one of those forms of damage that will get a different answer from every painter who is asked the best way to repair them. Most refinishers will generally agree that the best way to remove a stone chip is to sand down to the first undamaged layer, which could be the primer, or even down to the parent substrate. When sanding down, it is very important to consider the feathering out of the paint layers. Each coat should be visible as a wide band of colour around the stone chip; the wider each band is, the more gradual the dip from the top coat down to the base of the damage. Too sharp a dip will cause a defect known as contouring, which shows the edge of each layer of paint like the contour lines of a map. Once paint has been feathered, it needs to be isolated with a primer to prevent mapping, which is when new coats of paint reactivate the edges of the old coats and make them wrinkle. This shows the outline of the repair. One way to prevent mapping is to fill the stone chip with a stopper, and then sand the stopper level to the surface of the paint. In theory this will not require any form of primer. The truth is that there is a downside to this type of repair, which is that the edge of the stone chip is visible partly due to stopper being porous and partly due to the edge of the stone chip being heavily fractured at a minuscule level which is easily seen in Figure 2.2. Realistically, if a panel such as a bonnet or front bumper has been brought into the workshop for stone chip repairs, there are usually concentrated areas which will definitely need taking back to sound substrate before etch primer and primer surfacer are applied. Once the concentrated damage is feathered out, any odd stone chips further back on the panel can be stoppered. As an aside, if one does decide to use stopper to fill a stone chip, the thinnest skim should be applied, and this should then be leveled with a block. Strictly speaking, fillers and

Figure 2.2 Stone chips. The impact of the stone raises the edges of the repair. Here there are loose flakes of paint clearly visible at the top edge of the chip.

stoppers should not be wet flatted; however, in this instance, the use of water to lubricate the abrasive is of great benefit to the technician. Of course, if a stone chip has started to rust, it is vitally important to sand back to bare steel and trace the spider rust (proper name filigree) back till it is fully visible. It is recommended to treat this with a rust converter; however, very often it is possible to sand the rust completely off the steel. It is most important to use an etch primer with a resistance to corrosion to apply on the bare metal area in order to prevent further rusting.

2.1.2 Scratches

Similarly to stone chips, scratches usually require feathering out with a dual action (DA) sander; however, they can sometimes be light enough to be flatted with a fine-grade abrasive and polished although the one in Figure 2.3 is far too deep for this. It is not advisable to try to fill scratches with a stopper as has been mentioned with the stone chips, as the author has witnessed several people try to do this, and nine out of ten times it has failed and wouldn't have taken much longer to do it properly. The key point with sanding scratches back is to only sand as far as required. It may be possible to sand a scratch down as far as the primer and then recoat with water-borne base coat which will not react with previous solvent base coats; therefore, it is important to establish if the panel has been painted with waterborne coatings previously. As vehicle manufacturers are increasingly using waterborne coatings from the factory, it is getting more and more likely that this is what will be over coated; therefore, it is still recommended to prime before applying colour.

Figure 2.3 Note the jagged edge of the scratch where the paint is shattered as the scratch is made.

2.1.3 Scuffs

It can be argued that scuffs are merely a concentration of many deep scratches in one area like those seen in Figure 2.4. They are usually a fairly severe form of paint damage and can also contain some panel deformation. Either way, scuffs require sanding back to the sound substrate and feathering out around the edge of the repair. If there is panel deformation, this will require filler for reshaping purposes, and again in both cases, a high-build primer will be required with an adhesion promotor on any bare substrate.

2.1.4 Small dents

The size of a small dent is, of course, subject to opinion and conjecture, but in this instance, we'll consider a small dent to be something less than the size of the palm of one's hand and shallower than 3–5 mm like that pictured in Figure 2.5. Most dents can be improved with panel beating techniques such as planishing or panel pulling. However, as this book is about refinishing, we'll only consider the small dents that can be satisfactorily repaired using only filler. Preparing panels for filler will be discussed in Chapter 3, but needless to say, the panel should first be cleaned with a solvent panel wipe to remove any contamination prior to sanding, and then prepped in accordance to the technical information relating to the filler being used. Some filler manufacturers indicate that their products can be applied over well-sanded original equipment manufacturer paint, whereas others stipulate that their product must be applied over bare metal/plastic. Once the panel is satisfactorily prepared, the filler can be applied and cured as necessary, before being shaped and primed.

Figure 2.4 This vehicle has scuffs on the plastic bumper, which are likely caused by hitting masonry.

Figure 2.5 This small dent is in a very unusual place. The edge of the pillar needs planishing back prior to filling the dent.

2.1.5 Corrosion

First, let us discuss the use of the word corrosion. The reason for using 'corrosion' rather than 'rust' is that only ferrous metals rust, yet a refinisher will regularly come across aluminium corrosion too. Corrosion that has been allowed to perforate the panel will need to be cut out and repaired by panel beaters. Corrosion that has yet to perforate the metal should have any remaining paint (Figure 2.6) sanded off entirely, then as much of the corrosion (of which there may be plenty like in Figure 2.7 should be sanded off too, before the corrosion is treated with a suitable rust converter. Once the corrosion is neutralised, it should be skimmed with filler if there is any remaining pitting.

2.1.6 Replacement panels

If the damage is so severe that the panel will need to be replaced entirely, it will be necessary to prepare the new panel correctly. As has already been discussed in Chapter 1, metal panels are usually supplied in e-coat as in Figure 2.8. Depending on the recommendations of the paint manufacturer, the e-coat may simply require thorough cleaning. However, some technicians prefer the reassurance of mechanical adhesion too, which is often done with a grinding fleece ('Scotch-Brite') or flatting with a DA sander. When using a DA to flat e-coat, it is recommended to use an interface pad between the backing pad and the DA disc. The interface pad helps prevent exposing bare metal on edges and swages, and it also aids the DA disc in contacting the curve of the panel more uniformly. The grade of the abrasive is left to the discretion of the technician; however, any grade between P240 and P400 is acceptable in order to provide mechanical adhesion for wet-on-wet primers, and in the experience

Figure 2.6 This corrosion is under the paint. However, it needs to be taken back to bare metal to fix properly.

Figure 2.7 This corrosion is far more obvious. However, once the paint is stripped back, the repair will be similar to that in Figure 2.6.

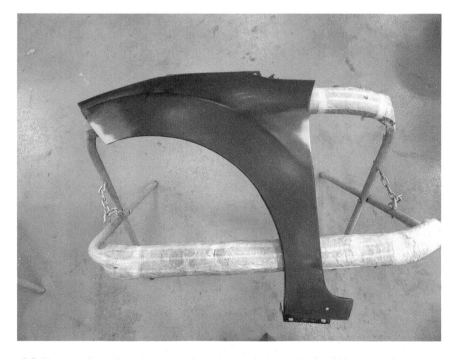

Figure 2.8 E-coat will need sanding smooth, and any rub throughs should be primed.

of the author, most technicians prefer P320. If when flatting e-coat, the parent material is showing through either due to accidentally rubbing through or due to necessity to remove a scratch, the bare metal should have an etch primer applied (Figure 2.9) before the wet-on-wet primer is applied. If the replacement panel is plastic, it can be prepped in a very similar manner, but the etch primer should be swapped for plastic adhesion promotor.

While identifying the damage, it should be considered how the repair will be undertaken. This is where there is a lot of strange terminology, such as edge-to-edge, blend and fade-out. Often a repair will necessitate a combination of these different paint techniques. For instance, a new panel such as a wing or door will require an edge-to-edge repaint, that is where the entire component has fresh colour applied from one edge to the other (and especially the edges). As the new paint is very unlikely to match the previous paint, it is usual to perform a blend repair on the panels' either side of the edge-to-edge panel. The exception to this is often bumpers such as the one in Figure 2.10, commercial vehicle panels and when applying direct gloss paints.

Most repairs include fade-outs and blends. The idea of these is to apply fresh paint on an undamaged panel on the edge adjacent to the new or repaired panel. The fresh colour is then 'flicked out' through the panel to give a graduated blend from the new paint to the existing finish. This can be clearly seen in Figure 2.11. This is then sealed in using a clear coat to provide a smooth glossy consistent finish. Blends can also be kept within a panel if the damage is far enough from either side of it. For instance, this may be stone chips on the leading edge of a bonnet or small dents or scratches in the middle of a door or wing.

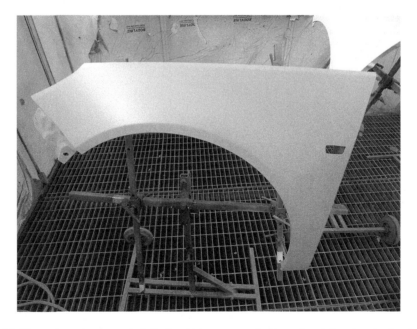

Figure 2.9 Once the wing is sanded, it should have any rub throughs etch primed prior to non-sanding primer being applied. It is not usually necessary to apply a full coat of etch primer like this wing has. However, some paint manufacturers suggest one or even two heavy coats like this to help with corrosion resistance.

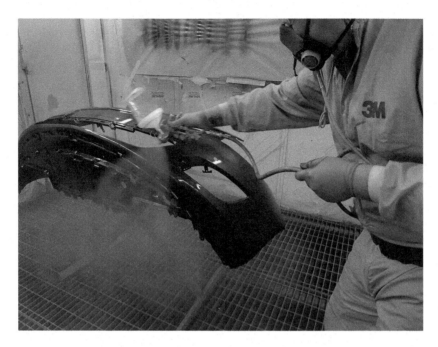

Figure 2.10 Most new panels have the fresh paint blended into the adjacent panels; one of the exceptions is bumpers, which can be repaired edge-to-edge.

Figure 2.11 When studying the masking in this picture, it is quite easy to see where the colour fades from deep blue on the quarter panel through the back edge of the door, and then gets lighter as it blends into the old paint through the length of the door.

When examining the panels adjacent to the repair panel, it is very important to refer to the job card, because if there is any unrelated damage on the blend panels, it will be recorded, whether or not this should be repaired or simply ignored. If it is to be repaired, this is known as 'betterment'. Betterment may be offered to a customer as an extra that they pay for, or sometimes as a sweetener to help gain the job in the first place. If the damage is to be ignored, this should be confirmed with the supervisor, and the blend will then be performed on the panel over the damage, as in Figure 2.12, and the panel finally clear coated.

Figure 2.12 The chip evident in the picture has been painted over during a previous repair. This may have been due to the adjacent panel having a repair very close to the edge of the door. This will have required blending into the next panel. The technician should have applied a small amount of stopper to this chip to create a better finish.

Chapter 3

Application of foundation materials

Before we discuss how to apply foundation materials, perhaps we should first ask, 'What exactly is a foundation material?' Like any building, a good paint job requires a strong and stable foundation. Also, like a building's foundations, foundation coats shouldn't be seen. Therefore, a foundation coat is any refinishing material that is applied to a repair, but is not the final colour or clear coat. Some are designed to level or smooth out a repair such as the filler featured in Figure 3.1, some to provide adhesion, and others to provide corrosion protection – either through isolation from moisture or through chemical protection. A huge selection of foundation materials are available, and here we'll cover the most commonly used, which handily also covers nearly all foundation materials. The intention is to list the correct use and compatibility of each of the types of foundation materials in the same order as they would be used on a vehicle. This should provide a logical sequence of use. Weld-through primer

Figure 3.1 Foundation materials are an important part of refinishing repairs.

could be listed as the first foundation material, as it provides inter-panel moisture protection. However, it is usually considered to be something used by panel beaters rather than refinishers. For obvious reasons, only foundation materials applied in body shops will be listed, so e-coat will not be included in this chapter, even though it is arguably a foundation material. It makes sense to start with fillers, although again these are often used by panel beaters too.

3.1 FILLERS, STOPPERS (2K AND 1K) AND GLAZES

3.1.1 Fillers

Fillers are sometimes given a bad name and considered a 'bodge' – this is mainly due to misuse, and expecting too much from them. If the correct filler for the job is selected and it is applied in modest coats, it is certainly not a bodge but a necessary material used to enable repairs to be completed economically. On the topic of the modest use of filler, this is subject to the opinion of the individual technician. However, as a rule of thumb, a single coat should not really be more than 5 mm thick, and a total fill should have a maximum of 7–10 mm. Having said that, the thinner the total fill the better. Furthermore, there are some repairs with only a few millimeters of filler which have cracked in use, and other vehicles have driven around for years with 20 mm or more which have not cracked at all. Fillers are applied with a filler spreader (Figure 3.2). These have a flat side and a side with three chamfers. Strangely, in the authors' experience in the refinishing and repair trade, these two sides cause more arguments than anything else. One camp insists that the chamfers (Figure 3.3) should be used to help the technician angle the spreader against the filler and panel. The other camp insists that the entirely flat side (Figure 3.4) should be used to apply the filler, and the chamfers are merely used to provide flex in the spreader. There is a third group who don't care which side of the spreader they use, but these technicians are confused and shouldn't be trusted! Lastly, you have the monsters who don't take any notice and will happily use the fourth side that has no chamfer (Figure 3.5). Don't be a monster. In case you're wondering, the author is very much for the chamfer side being used on the filler. However, in conversation with a major filler manufacturer, it would appear that they consider the chamfer side is simply there to aid the removal of the spreader from the mould! When holding the spreader, it is easy to forget that they have a long side and a short side. Depending on the amount of filler required, either the long or the short may be utilised. Despite this statement, a useful point of using the wide side is that the spreader can be flexed over one's thumb, as shown in Figure 3.6, thus helping to provide a feather edge in the newly applied filler.

All two-pack fillers require a hardener, more often than not benzyl peroxide (Figure 3.7). This should be thoroughly mixed in, on a nonporous board; some technicians will use plastic mixing pallets or old bits of steel or aluminium. Many filler and paint manufacturers make a dedicated mixing pad which consists of a stack of nonabsorbent paper bonded on three sides and loose on the fourth. This fourth side lets the operator peel off a soiled sheet to expose a fresh clean sheet Figure 3.8. This is the author's favourite tool for the job due to the quick cleanup. Mixing on cardboard should be avoided as the porous material will suck the resins out of the filler and make

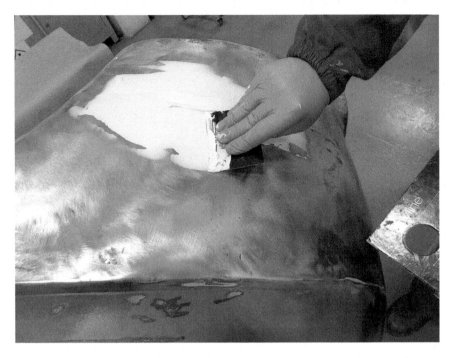

Figure 3.2 Applying filler with the wide side of the spreader.

Figure 3.3 Chamfered side of filler spreader.

Figure 3.4 Flat sides of filler spreader.

Figure 3.5 Do not use the edge with writing on to apply filler.

Figure 3.6 Flex spreaders to help feather in freshly applied filler.

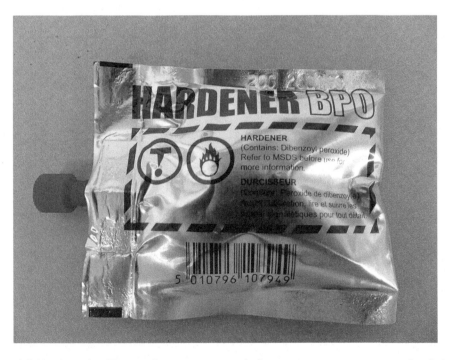

Figure 3.7 Hardener for filler usually contains peroxide. It is important to protect your hands from the peroxide with nitrile gloves.

Figure 3.8 Filler mixing pads provide a clean surface for each mix.

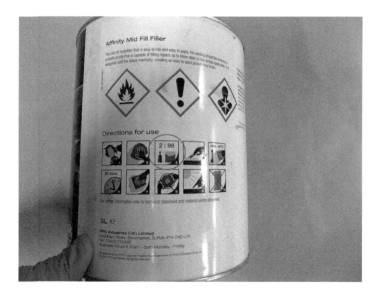

Figure 3.9 Most filler tins have the required mixing ratio on the rear. In this case, there is a pictogram of the tube of hardener with a 2 next to it. Therefore the filler manufacturer recommends 2% hardener is added to the quantity of filler.

it stodgy and more difficult to apply. When mixing fillers, the mixing ratio is generally between 1.5% and 3%. Technical data sheets (TDS) should be referred to, to ensure that correct mixing is undertaken, but nearly all fillers also have the ratio printed on the side of the tin (Figure 3.9). With practice, a technician will work out what the correct ratio is by eye and generally won't use the following method. This technique is a useful guide to be undertaken by new technicians, but more experienced ones may want to try this occasionally just to remind themselves what percentages look like and what the resultant colour of the activated filler should look like.

The technique:

- Once one has chosen the required quantity of filler, it should be arranged in as equal a circle as possible. This is 100% of the required filler (Figure 3.10).
- Cut this circle in half. This is 50% of the filler (Figure 3.11).
- Cut the semicircle in half. This is 25% (Figure 3.12).
- Half of 25% is 12.5% (Figure 3.13).
- Half of 12.5% is 6.25% (Figure 3.14).
- Half is 6.25%, which can be rounded down to 3%. Generally, this is the maximum amount of hardener required (Figure 3.15).
- This could also be cut in half to show the minimum amount of hardener (1.5%).
- Figure 3.16 shows the resultant tiny portion of filler, match it in size with the correct hardener.

Figure 3.17 shows how the little blob of filler should then be mixed carefully with the original amount of the filler. Figure 3.18 shows the reason for the hardener being a contrasting colour as it is easy to see when it has not yet been fully mixed. Figure 3.19 shows what the filler should look like when fully mixed. When mixing, it is good practice to push down hard on the filler with the spreader as it is mixed, to squash out any air bubbles that may show as pin holes later when the filler is cured and rubbed down. On the subject of curing filler, a nice slow cure time of 10–15 minutes is preferential to a quick cure time. A slow cure will allow any solvents present in

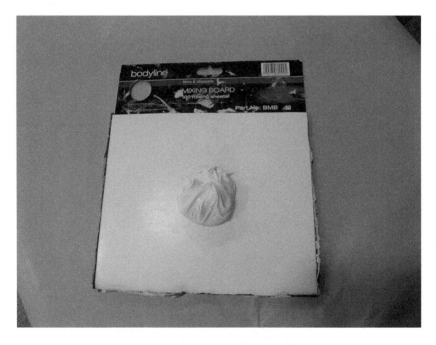

Figure 3.10 This is all of the filler to be used. Therefore, it is 100% of the quantity.

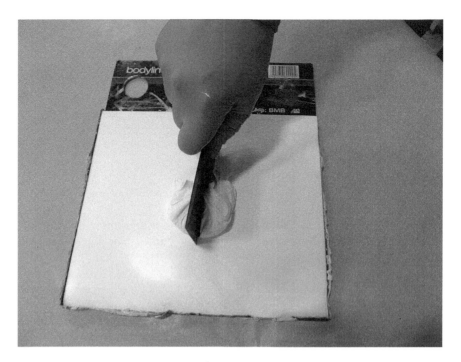

Figure 3.11 Cut the filler in half to gain 50% of the total quantity of filler.

Figure 3.12 Cut the filler in half again to get 25%. Note the rest of the bulk of filler on the corner of the mixing pad.

Figure 3.13 Cut the 25% of filler in half to gain 12.5%.

Figure 3.14 Cut this quantity of filler in half to gain 6.25%.

Figure 3.15 Once the 6.25% of filler is cut in half, it will leave 3% filler.

Figure 3.16 As we want 2% hardener, we dispense a neat pile slightly smaller than the 3% blob of filler.

Figure 3.17 Mix hardener in with all of the filler.

Figure 3.18 There should be no streaks of hardener left in the filler. If it looks like this, continue to mix.

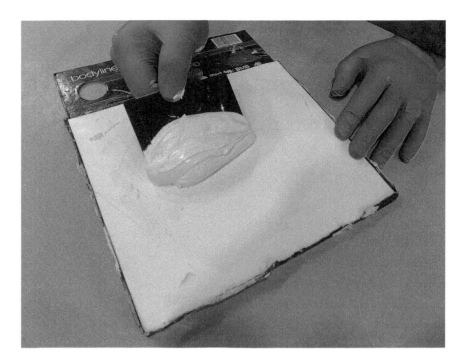

Figure 3.19 This filler is fully mixed and ready to apply.

the filler to evaporate and give the filler time to properly polymerise. This is why it is important to observe the recommended quantity of hardener for the filler being used. Too fast a cure, apart from potentially causing the filler to go off before it can be properly applied and shaped, can also cause the filler to cure in a brittle, crystalline way. However, when mixing it is best to err on the side of the faster cure as if it is slightly too hot a mix, it will be fine, but the author has witnessed many people mixing filler with too little hardener. This ends up not going off and is ultimately scraped off the panel into the bin. Remember if there are any streaks of colour in the filler, it is not fully mixed. When applying filler it is a good idea to feather out the edges as in Figure 3.20, and make it as neat as possible as in Figure 3.21. After application don't forget to clean up the spreader! Figures 3.22 and 3.23 show a good before and after, remember if the spreader still has filler on it it's not clean!

3.1.1.1 Bridging fillers

Bridging fillers are not often used, but some panel beaters prefer to use them if there is a very deep fill to undertake. This is due to them having virtually no shrinkage, especially with the carbon fibre-based ones, and these also provide excellent flexibility – an important factor on larger panels. Some would argue that if there is a job that requires such a deep fill, it probably needs further panel repair first. These fillers should never be used to bridge rust holes, as over time the filler will simply fall out as the corrosion continues to develop under the filler material. This is of course true of all filler materials. Most bridging fillers are based on glass fibre strands but some are

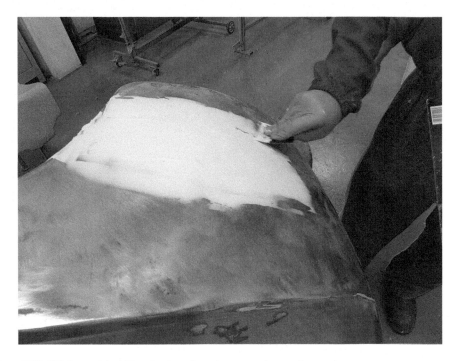

Figure 3.20 While applying filler, feather the edges to make sanding easier.

Figure 3.21 Make filler as neat as possible before it starts tearing. The graduation at the edge of the filler will help it blend fully with the contours of the panel once it is sanded to shape.

Figure 3.22 Use a solvent panel wipe to clean filler from spreaders. Thinners can be so aggressive as to damage the plastic spreader.

Figure 3.23 With a wipe of paper towel, the spreader is clean.

available with carbon fibre-reinforcing strands instead. Bridging fillers are especially useful for filling damage in composite components such as glass reinforced plastic (GRP) bodies and styling kits.

3.1.1.2 Deep-fill fillers

Deep-fill or lightweight fillers are usually polyester based and the thickening agent is quite often talc (from which talcum powder is made?). Generally speaking, a deep-fill filler is whipped up with air like whipped cream during the manufacturing process. This has the threefold benefit of making the filler lighter in weight, more flexible when cured and easier to sand into shape. Unfortunately, the resultant air bubbles have the drawback of leaving many pinholes in the filler, which will require further filling with stopper or glaze. Some manufacturers add hollow beads into their lightweight filler mix – this again decreases the weight of the filler with no detriment to the strength.

There are dedicated flexible fillers (Figures 3.24 and 3.25) for use on plastic components such as bumpers. Standard fillers often provide an adequate level of flexibility; therefore, the fillers dedicated for plastic components are not often used. Preparation of foundation materials ready for top coats will be discussed fully in Chapter 4; the process of flatting fillers is examined in this chapter.

3.1.1.3 Medium-fill fillers

As the name implies, these fillers are not designed to be used for severe panel deformation. They are made to be used for light dents, and as they do not need to be used for

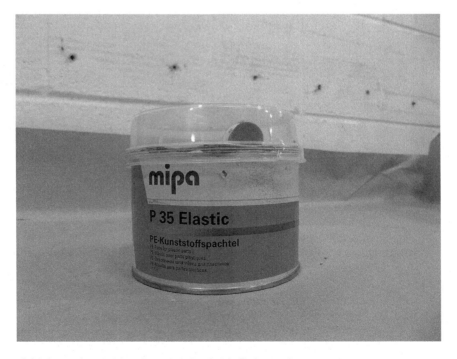

Figure 3.24 Filler intended for plastics is clearly labelled as such.

Figure 3.25 This vintage steering wheel has been repaired with two-part glue, then finished with a plastic filler. It can be seen in the unsanded state in the centre and prepped for primer on the rim.

such a thick coat, they do not need to be whipped up in the same way to make them lighter. A useful effect of not being whipped up is that they are less likely to need a coat of stopper for pinholes, and as they are applied in lighter coats, they do not need to achieve the same kind of flexibility. They instead rely on being used in thin coats to provide the flexibility.

3.1.1.4 Shaping filler

First, all fillers should be sanded dry – this is due to fillers being porous, so if they are flatted with water, they will soak it up, and it will become trapped under the fillers, causing corrosion and adhesion problems. The hygroscopic nature of body filler also means it is not recommended to use panel wipe to clean bare filler. The author has witnessed technicians soaking bare filler with panel wipe to gain a reflective gloss, as this shows waves in panels and panel shape exceptionally well. Panel wipe also has the potential to evaporate through high-build primer in the baking process, causing blistering and Solvent Pop. When shaping filler, it is recommended to use an extracted block (Figure 3.26) rather than a dual action sander, as the block can be easily manoeuvered into angles that will help regain panel contours.

A dry dust guide coat (Figure 3.27) provides an excellent visual aid to help show where one has sanded, low spots and pin holes (Figure 3.28), and it is recommended that a dust guide coat be used repeatedly, as it will show any wild scratches left from the blocking process.

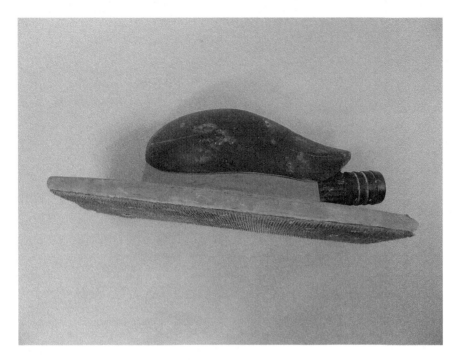

Figure 3.26 Extracted dry flatting block.

Figure 3.27 This dust guide coat is applied by squeezing the handle of the applicator while wiping the panel with the sponge. There is a bottle of dust in the middle of the applicator.

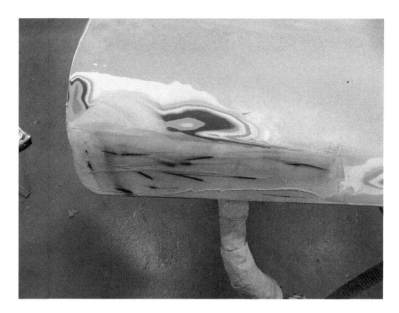

Figure 3.28 This filler has had a dry guide coat applied to show high and low spots while sanding.

Usually, the shaping as shown in Figure 3.29 is undertaken with an 80-grit abrasive paper first, as this provides fast removal of unwanted material, but it isn't too coarse.

At this point, it is often required to apply a second coat of filler. Once that is shaped, it is typically necessary and in fact recommended by many manufacturers to apply a stopper or glaze over the shaped filler.

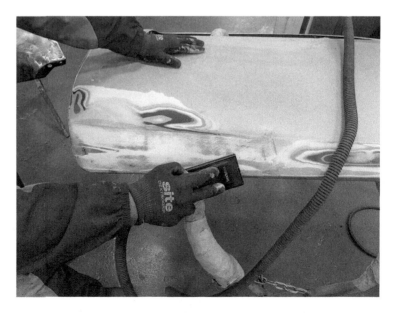

Figure 3.29 Here the filler is being shaped with P80 on a block. Note the high spots are white and the low spots show the guide coat still.

3.2 STOPPERS (2K AND IK)

Stoppers are filler materials that are used to fill pinholes and sanding scratches as well as very slight undulations in a panel. They were traditionally single-pack. Single-pack stoppers had several weaknesses, including slow drying times, and the application had to be in extremely thin skims. They often required wet flatting, which is far less than desirable on fillers. Single-pack stoppers are generally made from a form of very thick paint, originally cellulose, but now most often acrylic. Having listed the drawbacks, it should be noted that a good-quality single-pack filler is very useful for dealing with slight sanding scratches and minor pin holes, especially if they have been missed and the vehicle is about to be painted. It may be argued that this is a 'bodge', but if applied correctly, and the technician knows the limitations of the product, it can really get you out of trouble. Indeed Figure 3.30 and 3.31 show a very nicely executed repair done with single part stopper, chosen for ease of use and the fact that these slight scratches were well within its limitations.

Most stoppers used today are two-pack and are basically very dense fillers. This gives the advantages of a more stable foundation material than single-pack stoppers, less shrinkage, faster drying times and dry sanding compatibility. However, even with two-pack stoppers, their limitations should always be remembered. They are not intended to fill dents, as their density makes them inflexible, and second, this density means they can be harder to sand than lightweight fillers. Therefore, it is tempting to start with an 80-grit abrasive paper. If a high-build primer is applied over filler with 80-grit scratches still present, they WILL show through. This may happen

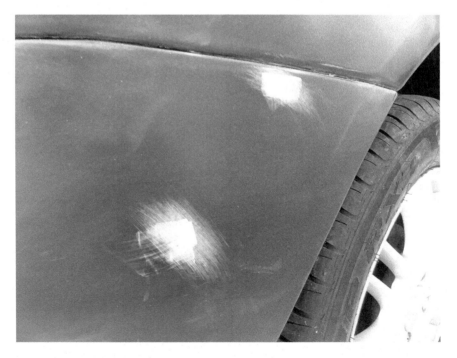

Figure 3.30 The small chips in this bumper are keyed with P320; then a single-pack stopper is applied. This will be sanded with P800 wet and dry.

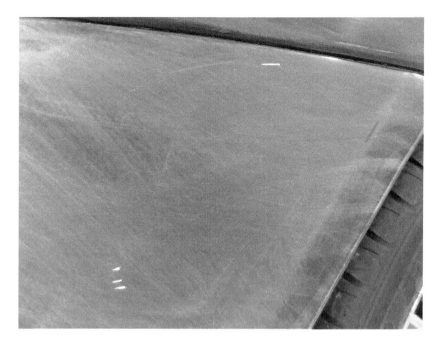

Figure 3.31 The chips are now filled with stopper which has been sanded level with the surface. The fine scratch marks of the P800 are easily covered with a rub-through primer on the stopper.

immediately or months after the repair has been undertaken, but the high-build primer will eventually sink into the scratches and show in the final finish. Assuming one is now happy with the general shape, the final shaping should now be undertaken with 120–180-grit abrasive paper. At this point, high-build primer can be applied. Usually, high-build will successfully fill 180-grit scratches, yet there is a risk that it can still sink into these scratches and show through the final repair. Many repairers will use 320 grit to provide the final finish on the filler prior to the application of high-build primer as this provides a scratch-free finish. Although this is standard practice, the author has observed the primer peeling from such a substrate on several occasions. It is therefore recommended that 240 grit be used as an excellent compromise for providing mechanical adhesion, and yet give fine enough scratches to be filled with high-build primer.

3.3 GLAZES

Glazes (often referred to as self-leveling putties) may be classified as a type of 2K stopper. Figure 3.32 shows how runny self levelling compounds are. The advantage is their 'runniness', which, as the name implies, helps the material level out when used on a horizontal surface. They can be quite difficult to apply, but those technicians who like glazes can apply an even, thin coat over a high-build filler in such a way that minimal sanding is required in order to gain a fine pinhole-free surface.

Figure 3.32 Glaze is often called self-leveling putty. As can be seen in the picture, it is very runny. This can make it tricky to apply on a vertical surface.

3.4 PRIMERS

3.4.1 Etch primers

Etch primers (Figure 3.33) are, as the name implies, designed to etch into the substrate. They are principally designed for use on bare metals, and due to their acid content, promote adhesion. They are not recommended for applying to plastics, this also includes filler. Application to repaired panels should therefore be undertaken carefully, applying etch only to the bare metal areas and not the filler. Traditionally, etch primers contained chromates, which had excellent anticorrosion properties, but unfortunately they were not so good for the refinishers' health, so they have been largely removed from the ingredients lists of etch primers. Etch primers are often referred to as 'wash' etch and self-etching primers, and they are usually applied in a single, very light coat. However, the author has spoken to one technical trainer who recommends two coats to help make up for the lack of chromates. Now – a contentious point – is there a need to cover etch primer with a further primer before colour coats? There is no simple answer to this; the correct technique will depend on the size of the bare metal area, the location of the component on the vehicle and type of colour coats. Needless to say, missing out an important layer may result in a failure of the entire paint system, so if in doubt, use a wet-on-wet or high-build primer surfacer to overcoat the etch primer as required. If it is decided to overcoat an etch with a colour coat and no further primer, it should be made sure that the bare metal is entirely coated first. The etch can then either be overcoated immediately, or if it has been allowed to dry, it should be abraded first.

Figure 3.33 Etch primer requires its own thinners. This usually contains the acid etching component of the coating.

3.4.2 Adhesion promoters

Sometimes referred to as plastic primers, adhesion promoters are designed to help the above coats of paint stick to plastic components. Etch primers are also adhesion promoters, but they are treated separately, and we have already covered them. It is generally accepted that the term 'adhesion promoters' refers to those for thermoplastics. Adhesion promoters (Figure 3.34) are only required on bare plastic and should be used wherever an original equipment manufacturer (OEM) finish has been sanded through to bare plastic, as well as on brand new components. Regarding new plastic components, it is usual for them to be supplied with a primer on them; however, it isn't always obvious whether a component is bare plastic or primed. Therefore, it is recommended to sand a small area with a fine-grade abrasive such as P800–P1000 and check the colour of the dust or sludge. Primer will sand much easier than plastic, and this should be very evident with the quantity of dust from a few seconds' sanding. Adhesion promoters are often delivered from an aerosol (Figure 3.35) as they are single-pack, so they require no mixing, and aerosols offer the convenience of being ready for use with very quick preparation. Adhesion promoter helps further coats stick by modifying the surface of the plastic component. It is often completely clear in colour and should be applied in a single light coat. Depending on the manufacturer, it is applied sometimes a few minutes prior to priming or maybe 15–20 minutes prior to over-coating. As well as the completely clear type of adhesion promoter (which resembles thinners), there are some with a silver tinter, which helps the user to see which parts are covered. There are a few true plastic primers. These are adhesion promoters with a degree of build to them. These plastic primers, like the clear thinners type,

Figure 3.34 This type of plastic primer is poured into a spray gun of around 1.4 setup. It is useful when repairing larger areas of bare plastic.

Figure 3.35 Aerosol adhesion promotor is very handy for smaller repairs and plastic rub-through primers as it is ready for use.

are available in a tinned format suitable for applying via spray gun. Finally, adhesion promoter wipes are available – these resemble wet wipes. Adhesion promoter wipes are single use, their main advantage being that they can be used very precisely with no overspray or masking necessary. Once the adhesion promoter is applied, a non-sanding or high-build primer should be applied prior to colour coats.

3.4.3 Wet-on-wet primer

Wet-on-wet primers are often referred to by paint manufacturers as non-sanding primers. Simply put, they are a primer that provides a modest level of build and isolation from the substrates below, and yet do not require sanding prior to recoating with colour coats – hence the term non-sanding, which means wet coat of paint on wet coat of paint. Depending on the paint manufacturer, wet-on-wet primers may be mixed in many ways. Some manufacturers produce a separate primer to be mixed with hardener and thinner, others will use the same primer as their high-build primer, but mixed with more hardener and more thinner to provide a lower viscosity. This lower viscosity is required in order to help the primer flow out into a smooth even coat (Figure 3.36). Manufacturers often use mixing ratios of between three and five parts primer to one part hardener. However, unlike many high-builds, there are usually

Figure 3.36 Wet-on-wet primer cures very quickly and is characterised by having a nice smooth finish with a semi-matt sheen. It can be de-nibbed after 15 min.

10%–30% thinners added too. In general, non-sanding primers are applied in a single wet coat very much like the last coat of clear coat or direct gloss paint. They need to be a wet coat in order to flow out properly with no orange peel, as this influences the way in which base coat, especially metallic and pearl base coats, will lay. In the author's experience of seeing many first attempts at applying wet-on-wet primers, they are one of the hardest refinishing materials to apply, because if they go on too dry and textured, they will need to be sanded smooth; therefore, they will need to be cured. If they are applied too wet, they will run and again require baking and sanding. This is a problem when a wet-on-wet primer is meant to be a time-and-labour-saving material. Non-sanding primers can usually be de-nibbed of any dirt particles after 15 min of drying. This means that most technicians will apply the wet-on-wet primer, then go and clean the spray gun and mix the base coat, giving the primer time to dry; then if it requires a de-nib, it is already dry. After de-nibbing, it should of course be cleaned with a tack rag to prevent dust contamination in the base coat. Many manufacturers of non-sanding primers are now recommending that, if their products are to be applied over a brand new e-coat panel, as long as there is no damage, it can simply be panel wiped, and then the non-sanding primer can be applied directly over the un-sanded e-coat.

3.4.4 High-build primers and surfacers

When discussing high-build primers, the first point to be made is that many technicians are using surfacers when they think they are using high-build primers (which are often referred to as primer-fillers). This statement is simply backed up while checking several manufacturers' TDSs for their high-build primers. They usually state that they should be mixed with only primer and hardener, and thinners are not mentioned. As soon as a technician decides to start adding 10%–20% thinner in order to help the primer lay smoother, they have inadvertently mixed primer-surfacer which has less build than a primer-filler. This is not a problem, apart from the fact that the paint manufacturers have given their recommended mixing ratios after years of research and endeavouring to meet the stringent volatile organic compounds (VOCs) criteria of many nations. It is therefore recommended that TDSs are strictly observed. When deciding on which primer to use, it ought to be remembered that primer-surfacers should only be applied over repairs with the slightest of unevenness – for instance, where a stone chip or key scratch has been feathered out. Primer-fillers like that in Figure 3.37, on the other hand, are designed to remove slight undulations in panels, such as where a filler repair has been undertaken; however, the emphasis should be on 'slight' undulation.

3.5 CHECKING THE QUALITY OF REPAIR

A technician should ensure that any repair is as smooth and well blended into the surrounding body work as possible, and should definitely not rely on a high-build primer to recreate panel contours. Figure 3.38 is a good example of what happens when the limits of wet-on-wet are pushed too far. The technician responsible for this should have checked their feathering better before priming with high-build. Most high-build primers are applied in three heavy coats, which give a deep build which,

Figure 3.37 High build is used to cover repairs and isolate rub-throughs.

Figure 3.38 Here the contours of the paint layers underneath can be seen through the wet-on-wet primer. This will need blocking down and priming with high build.

when flatted back, can be left thicker in the slight low spots and thinner in the higher areas. As has been previously mentioned, due to this thicker coating, high-build primers can be applied over repairs sanded with P180, although there is a risk in this situation of the primer sinking back into the scratches and show through final repairs, so P240 is recommended. High-build primers are available as single-pack (1K), usually in an aerosol format. These are considered by many to be inferior to two-pack materials, but they do offer quick, easy use if a small repair has to be undertaken. This needs to be done in a well-ventilated area, preferably the spray booth.

As always, the TDSs should be referred to, although it should be said that, as a primer-filler is expected to provide extra build, aerosol high build is often applied in three full coats. The majority of high-build repairs are undertaken with the use of 2K (two-pack) primers. These are most often acrylic-based materials, where the primer is mixed with an activator (hardener) to promote polymerisation (cross-linking). Polymerisation is what is taking place in 2K materials as they cure. As with all 2K materials, cure time is sped up by baking the wet primer either in a booth or with infrared heat lamps. Generally speaking, by heating two-pack primers, they can be forced to dry around nine times faster.

3.6 SPRAYING 2K PRIMERS

When applying 2K primers, it is important to choose the correct spray gun and gun setup. Due to the thicker viscosity of primer-filler (generally around 60 seconds in a DIN4 cup), it is recommended to use a slightly larger spray gun setup of around 1.8 mm, and as a primer-surfacer has a lower viscosity (generally around 24 seconds in a DIN4 cup), spray gun setups are usually around 1.6 mm like the gun pictured in Figure 3.39.

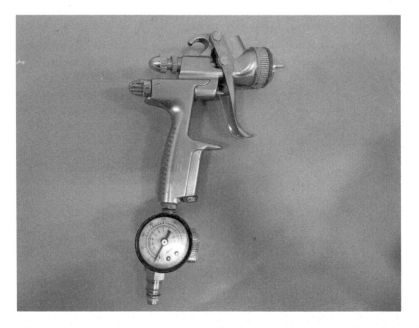

Figure 3.39 Primer guns can be of many different sizes. A 1.6 mm setup is typical for high-build application.

3.7 ROLL PRIMING

Some body shops opt to apply high-build primer with paint rollers. Paint rollers provide several advantages. First, the primer can be applied in the workshop, as there is no atomised paint mist. Second, there is less masking required as there is no overspray to worry about. The cleanup is also quicker, as the sponge section of the roller is simply thrown away. However, there are also some disadvantages. First, the primer dries with quite a rough texture – far worse than orange peel, and the resultant flatting can sometimes remove much of the primer. Second, if the first coat is not dry enough, the technician may peel the first coat off with the roller. Foam rollers and paint trays can be quite costly to replace continuously, although the trays can be lined with stat sheet to keep them clean.

3.8 EPOXY PRIMERS

Epoxy primers are gaining popularity in the United Kingdom, as they offer several advantages over traditional etch primers and acrylic high-build primers, in that they can do the job of both in one material. Epoxy primers are two-pack materials. They are often referred to as epoxy primer-fillers and, like etch primers, offer good anticorrosion properties. Also, as with acrylic primers, epoxy primers can be used as sealers or surfacers, as well as high-build primers. Having already mentioned that they are similar to etch primers in their anticorrosion properties, it is worth stating that they are also similar to etch primers in offering very good adhesion to a wide variety of substrates, including aluminium, galvanised steel and GRP. In addition, when used as wet-on-wet primers, they can dry very fast.

Restoration and custom workshops are particularly enjoying the benefits of epoxy primers for several reasons, for example, any vehicle body components that are taken back to bare steel can be quickly protected from corrosion with a coat of epoxy. This means that any panels that require no further work get a coat of corrosion-resistant primer that sticks extremely well. Also, once cured, the primer can be directly recoated with polyester stoppers and fillers, which means that any filler work doesn't require the panel to be taken back to bare metal first. Epoxy primers can be applied over all substrates already mentioned, plus OEM top coats and bare plastics. Further to these substrates Figure 3.40 shows a test to see if epoxy or etch and high-build stick better to sanded chrome. The important factor with any of the substrates is that they should be clean, dry and free of contamination. There, of course, should be no corrosion present, and the substrates should be sound (no loose or flaky material).

3.8.1 Applying epoxy primers

The same precautions should be taken when painting epoxy primers as with any other material, that is, using a spray booth and wearing suitable protective clothing and air-fed mask. Depending on the manufacturer, the spray gun setup could be anything from a 1.3 high volume low pressure (HVLP) or 1.4 compliant to 1.9 HVLP or 1.6 compliant, and the inlet pressure at the gun varies from 2 to 4 bar. Of course, these variables all depend on whether the epoxy is mixed as a wet-on-wet or high-build primer. After application, wet-on-wet and sealer primers take

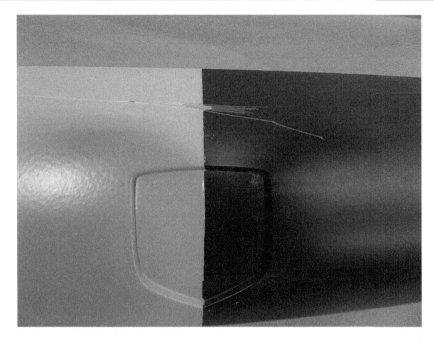

Figure 3.40 The grille piece pictured here was chrome the customer wanted it body colour. So an experiment was created. Epoxy primer is the light grey on the left. The dark grey is etch primer coated with high build.

around 15–20 minutes to dry at 20°C, whereas high-build primers generally take around 30 minutes at 60°C.

In conclusion, epoxy primers offer many advantages over traditional acrylic primers, but they are still used less often. This may be due to epoxy primers often costing more than acrylic primers, but it is also likely that technicians and workshop owners like to stick with what they know and are therefore being slow to move to epoxy primers.

3.9 POLYESTER PRIMERS

Polyester primers are rarely used in day-to-day crash repairs but serve a useful purpose in restoration and custom workshops. They offer very high film build qualities, which is a useful property when panels are not particularly straight. Polyester primers are basically extremely runny body fillers that are applied in one or two coats. Once dry, they can be sanded and shaped with coarse abrasives such as P120 or P180 on an extraction block. This provides an excellent leveling; however, they do have their limits, so the panel should be straightened and filled as best as possible before the application of polyester primer. It also offers the useful property that if the panel is still not quite straight, it is acceptable to apply a filler over the polyester primer. The drawbacks are that polyester primer is only reliably applied over bare steel, previously sanded filler, epoxy primer, or GRP. Many manufacturers recommend that it be applied over well-sanded OEM paint; however, in the authors' experience, this is not

100% reliable, and applying over etch primers should definitely be avoided as well as on thermoplastic or solvent-sensitive finishes.

As long as the above limitations are considered, the adhesion properties of a polyester primer on bare steel, epoxy primer and GRP are excellent. Depending on the manufacturer, one to four coats may be applied with a 10-minute flash-off between coats; however, two coats are normal. Whilst discussing application, it should be remarked that this is where polyester primer differs from most paints. The spray guns required have a very large setup – 2–3 mm is recommended by most manufacturers, with a gun delivery pressure of 3–4 bar being not uncommon.

Mixing ratios also seem unusual until you remember that this primer is actually a runny body filler; then the ratio of 1%–3% hardener in the entire quantity makes sense. Figure 3.41 shows an example of how polyester primers are often supplied with a litre tin of primer and a tiny bottle of hardener. Due to this small quantity of hardener, it is recommended to weigh the desired amount of primer and work out the weight of the required hardener, rather than trying to use volume on the side of a mixing cup. This does result in a very high viscosity primer, usually around 45 s in a DIN4 cup.

Drying times are not unlike acrylic high-build primers, being on average 30 min at 60°C. There are some technicians who prefer to leave polyester primer for a week before sanding, but consulting many TDSs has not brought this to light as a manufacturer recommendation. As has already been described, after the primer has cured, it can be sanded with abrasive as coarse as P80; however, P120–P180 provides fast shaping without the disadvantage of chasing 80-grit marks later down the line. Once the primer is correctly

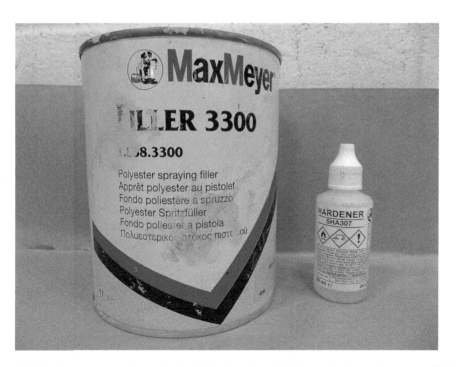

Figure 3.41 Polyester primer is supplied with tiny bottles of hardener. As the mix requires 1%–2% hardener.

shaped, P240 can be used to remove the P180 scratches. When the primer is prepared, it should be over-coated with wet-on-wet or high-build primer in order to seal it.

In conclusion, polyester primer has the useful properties of extremely high film build, which can be sanded and shaped to provide beautifully straight panels. Polyester primer adheres very well to certain substrates (bare steel, epoxy primer, GRP, body filler and well-prepared OEM paint) but reacts badly to application over etch primers, thermoplastics and solvent sensitive finishes. It is a foundation material that is not often required but can provide an excellent finish to vehicle panels that may otherwise have shown as 'wobbly' and obviously repaired after painting. In short, use it only when necessary and observe its slightly unusual properties, and mixing and application techniques.

3.10 SEAM SEALERS

Along with anti-stone chip, seam sealers are unusual in that they are applied with 'function over form'. This simply means that they are applied to do their job, and aesthetics are an afterthought. So what *is* a seam sealers job? Funnily enough, it is to seal seams! Seam sealers are also often referred to as body caulking; however, when technicians say this, they usually mean the extruded type.

Anywhere two or more pieces of metal are joined without a constant seam weld or adhesive layer; there is a risk that moisture will permeate at that seam and cause corrosion to form from the inside out as in Figure 3.42. The prime types of joints for this to happen are spot welded joints, which are numerous in vehicle body construction,

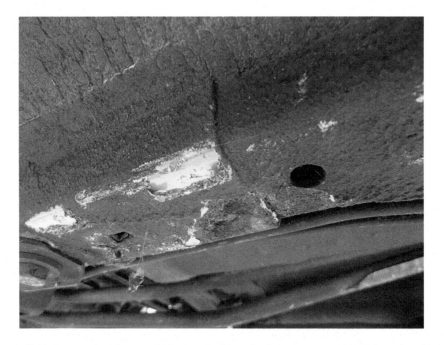

Figure 3.42 If this joint had been sealed properly from the factory, it would have been less likely to corrode so badly.

and clinched joints, such as those that are found around the edges of door, bonnet and tailgate seams. There are four main types of seam sealer used in vehicle refinishing: brushed, extruded, sprayed and taped. All of these perform the same function but will be used in different places and for different reasons. As with any refinishing procedure, when reapplying seam sealer to a repair area, the technician should try to replicate as close as possible the original finish, which occasionally means using a combination of two or more of the above types of sealer. Out of the four types of seam sealer, three (brushed, extruded and sprayed sealers) will be found on any OEM vehicle body shell. The last one, taped sealer, is used exclusively by the refinishing trade and is loved and loathed in equal quantity.

Sprayed sealer is the most awkward to try and replicate, and most manufacturers of seam sealer applicators provide different tips to match certain brands of vehicle. This will include instruction for delivery pressure in order to most closely match the OEM.

3.10.1 Sprayed seam sealer

Sprayed seam sealer is applied via an air-driven cartridge gun. As has already been mentioned, these sealer guns come with a number of tips which will provide stippled (Figure 3.43), ridged (Figure 3.44) and extruded seams.

Depending on the brand of sealer being used, it can sometimes be applied over bare metal, but more often sealer is applied over primed panels. As always, it is obviously important to refer to the TDS in order to ensure proper corrosion protection and durability. While referring to the TDS, it should also be consulted with regard to

Figure 3.43 Sprayed seam sealer.

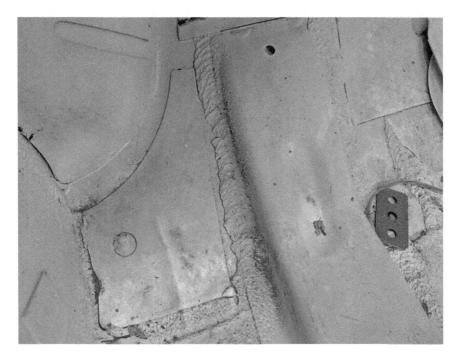

Figure 3.44 Ridged sprayed seam sealer.

drying times prior to top coating. Again, manufacturers vary their recommendations, but some can be used in the wet-on-wet technique. Sprayed seam sealer is most often used under floors and inside floor pans as well as boot floors, inside engine bays and inside wheel arches. It offers very thick, durable waterproofing of open joints and can bridge relatively large gaps. Of all the seam sealers, it is probably the messiest, so it is used in the least often seen places.

3.10.2 Brushed seam sealer

Brushed seam sealer is again used in the less seen areas, but compared to sprayed sealer, it is probably the easiest seam sealer to replicate and is extremely versatile, and the tools are obviously much simpler (a 25-mm, 1″-paint brush). The author finds it strangely reassuring to think that vehicles being produced now, like the mini pictured in Figure 3.45, are still being sealed by a person with a paint brush!

It is more normal to apply a brushed sealer over a primed metal, such as in Figure 3.46, and for the same reasons as listed in Section 3.10.1, the primer will provide adhesion for the sealer and further anticorrosion benefits. It is usual to let it dry for around 15 minutes prior to over coating with top coat.

3.10.3 Extruded seam sealer

Extruded seam sealer, such as the one shown in Figure 3.47, is the most often used sealer in the refinishing industry. It is mainly used for door skin replacement; however,

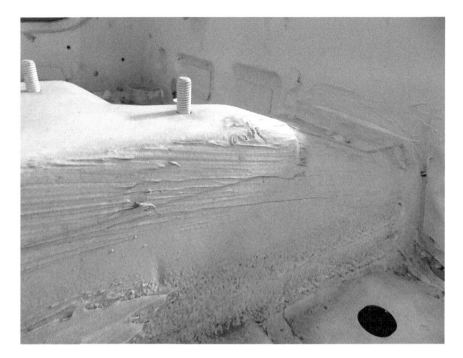

Figure 3.45 Brushed seam sealer found on the transmission tunnel of a Mini.

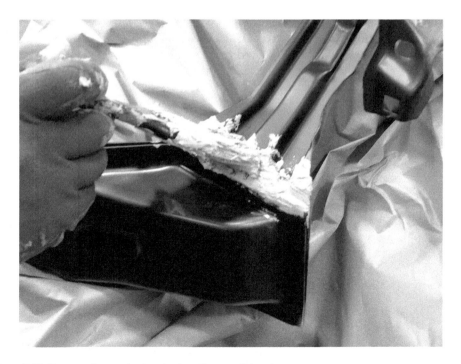

Figure 3.46 Seam sealer can be brushed easily onto joints but can get messy.

Figure 3.47 Extruded seam sealer on the back of a door skin.

it is also often used in visible seams such as van side paneling and around roof cant rails. When applying an extruded seam sealer, it is important to understand how to cut the nozzle in order to get the correct width and depth of bead. This can take a long time to master, and there is no hard and fast rule – many technicians have their routines that they will swear by.

As a general rule of thumb, the nozzle should be cut at around 45° at a point in its taper that matches the width of the seam (Figure 3.48). In order to gain a deeper seam, it may be necessary to cut a flat across the very tip of the nozzle (Figure 3.49). For windscreen bond, where a very deep fill is necessary, a V can be cut into the tip (Figure 3.50), but this is not usually used on body caulking.

On the subject of nozzles, it can be very irritating unblocking previously used nozzles, so a quick tip is to leave a short 'worm' of sealer sticking out of the end (Figure 3.51), which can be pulled out when the next person comes to use the nozzle.

When the panel is prepped, some technicians prefer to apply two strips of masking tape to set the width of the bead; if this works for you, then there is nothing wrong with it. Some technicians do not approve of this technique though, as when the masking tape is lifted, it can raise the edges of the bead and ruin the overall effect. Also, this technique only really works on straight lines, which means if it is applied to a curved seam, the edges of the tape are very likely to let the sealer bleed under which means the edges of the seam will not have the nice clean lines it should have. When applying the sealer, with or without tape, the seam should be used as the guide, and every effort should be taken to follow it in an accurate path. If the refinisher stops or changes the angle or speed while applying the caulk, it will affect the width and height of the bead, so it is important to plan the bead first. The technician should make sure

Figure 3.48 Cutting sealer nozzle to the correct width for the bead.

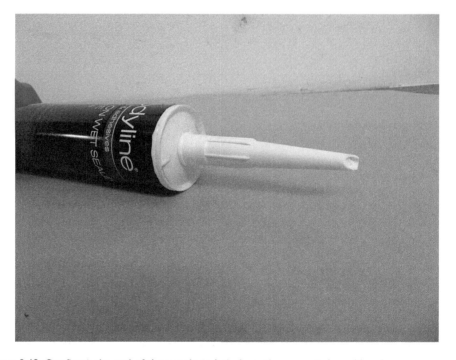

Figure 3.49 Cut flat at the end of the nozzle to help form the correct shaped bead.

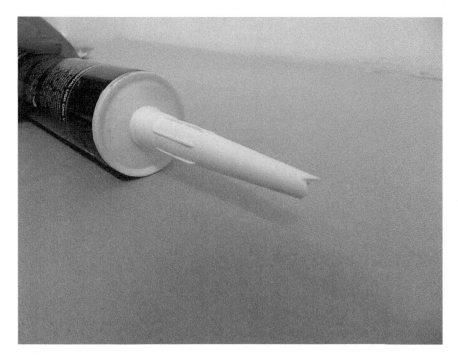

Figure 3.50 'V' shape at the end of the nozzle increases the depth of bead.

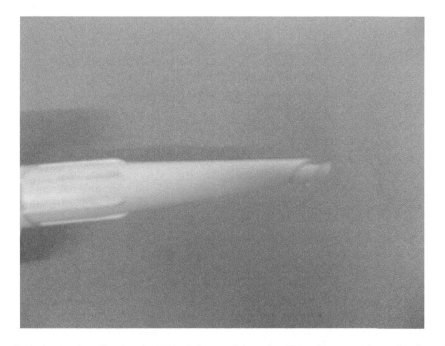

Figure 3.51 A quantity of sealer should be left out of the tube. This will cure and can then be pulled out by the next person.

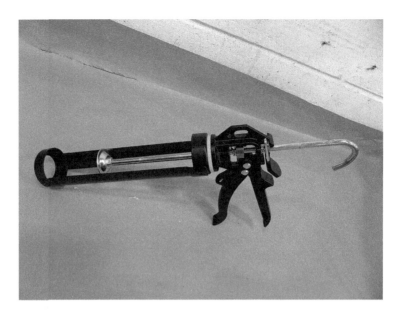

Figure 3.52 The manual sealer gun is cheap and reliable. But it does require the operative to trigger off and restart often in the middle of a seam.

that they can execute the pull in one long sweep, which may mean standing at one end of the panel and reaching to the far end. Of course, this is made more difficult once the cartridge gun's trigger is fully depressed, and one needs to open their hand and taken another squeeze. There are two alternatives (air-driven guns and battery-powered guns) that can help with this; however, we'll come to that later.

As has already been mentioned, many refinishers have their own techniques when applying caulking and will often use a solvent-based panel wipe to help smooth the sealer. This is frequently done with a gloved finger, but the author's preferred method is to use a soft paint brush dipped in panel wipe. Care should be taken not to over soak the caulk though, as this can have an undesirable effect on the structure of the cured sealer.

When choosing a sealer gun, there are many different designs. The most basic is the manual version, where the technician uses hand power to squeeze the trigger of the gun in order to push the sealer out via the mechanism (Figure 3.52). These guns have the advantage of being reliable, cheap and the least cumbersome, meaning they can be used in restricted areas. The alternatives are air-driven guns and battery-powered guns. The air-driven ones have been used for many years. They offer continuous beads via the depression of the trigger, which allows compressed air to push the ram through the cartridge. On the down side, they can get fouled internally if they are not cleaned properly, and the trailing air hose can mark panels, as well as upset the balance of the gun. The battery-powered electric types avoid this by being cordless, and they also offer continuous beads. They perform this by using an electric motor to drive the plunger via a screw thread; they do have the disadvantage of requiring charging and necessitating the technician to support the weight of the battery.

Most seam sealers are single-pack, drying simply through the evaporation of the solvent. However, some manufacturers offer two-pack sealers. These are usually self-leveling, which means they have lower viscosity than the single-pack type, and as with all two-pack materials, they also cure into a more durable substrate. This lower viscosity enables the refinisher to stipple, smooth, or ridge the sealer in order to replicate some of the sprayed effects. The drawback to this is that if a neat edge is required, it will be necessary to lay down the two strips of masking tape as described earlier.

3.10.4 Taped seam sealer

Taped seam sealer is the odd one out in this section, as it is only used in the refinishing trade as an alternative to caulking when replicating extruded sealer. At first glance, taped seam sealers (Figure 3.53) appear to be an excellent alternative to extruded caulking, as they can be applied in a very neat strip and are quick, clean and relatively easy to use. However, they aren't particularly popular with refinishers as they have a bad reputation for peeling away from the joint. This can be avoided, or the risk reduced, with some simple procedures which will be explained later.

Taped seam sealer is supplied on a roll with a strip of isolator plastic to stop it sticking to itself. Due to the sealer tape having to be a precut width, it can be quite difficult to match the sealer on the rest of the panels perfectly. It also has a very square profile, which is not too similar to extruded sealer. This is another reason for it not being particularly popular with technicians. Sealer tape is often clear, and if so, it is designed to be put on after the panel is painted, although some of these tapes are designed to be over painted, so it should be applied before the colour coats. It is important to check which type you are using. Now, as always, it is important to make sure that the joint is clean, dry and free of contamination before utilising the tape. It should be applied with a firm push and cut with a sharp knife at the end to give a nice clean line. Once in place, it should be fully seated with a small roller. Some technicians prefer a firm foam roller, and some a solid plastic one. Either way, the tape should have several passes applied to ensure that it fully adheres.

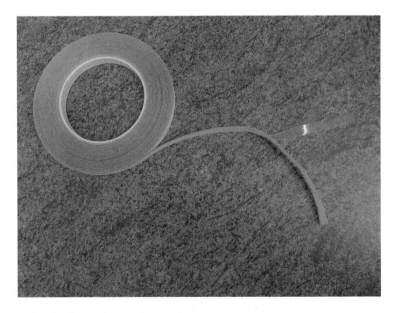

Figure 3.53 Taped seam sealer is usually applied over colour coats.

In conclusion, there are several types of sealer that can be used in repairing vehicle bodies. They all have their advantages and disadvantages, but the common factor is that they need to be durable, flexible and waterproof. The main deciding factor when choosing the type of sealer should be how well it matches the original sealer on the vehicle; one repair may indeed require two or even three types of sealer in a small area.

3.11 ISOLATORS

Isolators or sealers are intended to serve as a barrier layer that is applied over unstable or potentially reactive coats of paint. The affected paint may be unstable due to being old and oxidised, or perhaps inaccurately mixed during a previous repair. Numerous technicians have witnessed vehicle panels that have been painted many times prior to the repair they are endeavouring to undertake. It is often found in instances such as this, that any further coats will result in splitting, wrinkling, shrinking, or many other types of paint reaction.

The obvious solution to any of these cases is to entirely remove the affected paint from the bare substrate, but if this is not noticed during the estimating stage, there isn't usually enough profit to be made from the labour charge for a refinisher to mechanically or chemically strip the affected paint.

On most occasions, refinishers will choose to use wet-on-wet primers in order to isolate the new paint from the poor condition paint below. This is obviously only acceptable if the surface is smooth and free of scratches, chips, splits, or pinholes. On some occasions though, the wet-on-wet acrylic primer will react with the unstable paint below, thus magnifying the issue. In such situations, the alternative is to use a dedicated sealer product, such as the one in Figure 3.54. These are applied as a wet-on-wet primer, and they often take around an hour to dry. They are less likely to react with the unstable

Figure 3.54 Isolator coats are occasionally necessary when previous coatings are unstable.

paint below, as their chemical makeup involves a large quantity of methanol, a material not found in other types of paint.

A standard gravity-fed gun with a 1.3–1.4 setup should be used with a pressure of 1.8–2 bar. Having applied the isolating primer, it will need to be over-coated with a standard primer/surfacer. Due to the requirement of building a new paint finish on an unstable foundation, isolators are not often used, simply because most technicians would rather remove unstable paints and primers in order to expose a stable substrate. As has already been explained, this can take a long time to undertake; however, the refinisher can then proceed with the fresh paint with the peace of mind that the substrate is stable, and there can be no further faults from unstable paint. For these reasons, isolators are infrequently used.

3.12 ANTI-STONE CHIP

Along with seam sealers, anti-stone chip is another foundation material which definitely takes function over form. This is the main reason why it is used most often on the underside of vehicles (Figure 3.55) and in those areas which are both least seen and most vulnerable to stone chip damage, such as sills, valances and wheel arches. It achieves a notoriously unpleasant finish by possessing a very high viscosity when wet, thus causing it to land on the panel in thick blobs which then only flow a certain amount before drying.

Some refinishers like to add some thinners to the bottle of anti-stone chip in order to make it flow better, but this can affect its purpose – which is to dry with a very thick rubbery layer. It is therefore not advised to thin anti-stone chip down, unless it

Figure 3.55 Underbody coatings are usually a type of anti-stone chip and possess a thick rubbery property.

is to be used on the bottoms of doors or perhaps the leading edge of a bonnet, which several manufacturers have done in the past. When used in these areas, it is usually applied in a thin wet coat, allowed to dry, and then sanded back, although it is rare to find this as a standard feature in recent years.

Anti-stone chip provides protection by creating a cushioned layer under the top coat. This allows the top coat to flex when impacted by a stone or road grit. This prevents it from chipping, which will happen to the paint surface when it is impacted and has a hard substrate such as steel, or even thermoplastic underneath. In these instances, the paint is effectively smashed, which then results in the stone chip.

As has already been alluded to, anti-stone chip is most often found on the sills of vehicles, and here it provides excellent protection in an area which always suffers badly from corrosion. Depending on the manufacturer, anti-stone chip may be faded out where the sill shape transitions into vertical, or it may be terminated in a hard line somewhere on the vertical surface of the sill. The anti-stone chip finish can have smooth undulations, or a fairly sharp, pitted texture. Either way it is the job of the technician to match the texture of the original anti-stone chip as best they can, as well as matching how it terminates on the vertical contour of the sill.

It is often the case that not the entire sill will need to be recoated. In these situations, it is possible to fade out the new coat into the OEM anti-stone chip. Again, anti-stone chip is supplied in a plastic or metal bottle which is screwed to the bottom of an anti-stone chip gun. Anti-stone chip is usually supplied in a choice of black, grey, as in Figure 3.56, white, or beige, which will help with coverage by the top coat.

Figure 3.56 Anti-stone chip is applied at high pressure with a small fan pattern being achieved with the unusual suction-fed spray gun.

Figure 3.57 The anti-stonechip gun has a syphon tube long enough to reach the bottom of the bottles of anti-stonechip as supplied.

The anti-stone chip gun is effectively a very simple suction-fed spray gun (Figure 3.57), where air is passed over the top of a syphon tube, creating a vacuum which sucks the fluid up and mixes it internally with the air. This breaks the thick liquid up and deposits it in small blobs on the panel to be protected. Usually, anti-stone chip guns are connected directly to a compressed air system without a regulator. However, the use of a regulator can help with replicating the original anti-stone chip on the vehicle, as the size of 'blob' can be altered using different air pressures. This can be an advantage when trying to re-create a fairly smooth, or even a stippled finish. Anti-stone chip is usually a single-pack material; therefore, it can clean off the gun relatively easily, and any that isn't used in the repair can be saved in the bottle for the next repair.

Anti-stone chip may be applied over the top of the top coat as a contrasting colour or over the primer before being painted with top coat. The choice depends on matching the OEM anti-stone-chip on the vehicle, and whether the anti-stone chip being used is recommended by the manufacturer to be over-coated or not.

Finally, it is not advised to apply anti-stone chip direct to bare metal, and it is recommended to have the top coat either over or under the anti-stone chip to help create a waterproof layer in the film build.

3.13 RUB-THROUGH PRIMER

Rub-through primers are a fairly contentious foundation material to be included in a book of refinishing. Why? This is because in an ideal world a technician wouldn't flat through any paint layers when sanding panels. The reality is, however, that it is sometimes unavoidable to do so; therefore, when this happens, it is important to have a material that can be applied over the bare substrate in order to provide adhesion and protection. When a technician does rub though the paint or foundation layers on a panel, they usually want to repair their mistake as quickly as possible, and it makes sense then to use the ease and convenience of an aerosol primer.

There are many types of primer that can be considered to be rub-through primers, and the author has seen epoxy and etch aerosol primers (Figure 3.58) used successfully as well as cellulose primers used fairly unsuccessfully. Whatever primer is chosen, it should be applied in light coats. It has been witnessed on numerous occasions that when applied in a very wet coat, the etch primer will cause the damaged paint area to wrinkle up.

When preparing the aerosol, it should be warm (not too warm) to aid flow, and it should also be thoroughly shaken – again, to help flow, as the reducer should be properly mixed with the paint to gain the correct viscosity. Ideally, the chosen primer should be applied in light smooth coats which do not require sanding, as it is very likely that the same area will be broken through a second time if re-sanded.

When deciding whether to use a rub-through primer or to re-prime the area, there are several factors to consider. First, how large is the rub-through? Generally, they should only be a few millimeters wide, like the thin stripe of silver visible in Figure 3.59, but can sometimes be 40–50 mm long. Any greater than 5–6 mm wide and one should be considering properly priming the panel. Second, how stable is the paint layer that has been broken through? Remember that aerosols are single-pack and will have a lot of solvents in it, which will react with many paint edges, as can be seen in Figure 3.60 the aerosol primer goes on very wet. Third, how many

Figure 3.58 Rub-through primer is usually applied from an aerosol for ease of use. There is no mixing or clean up required so little labour time is wasted.

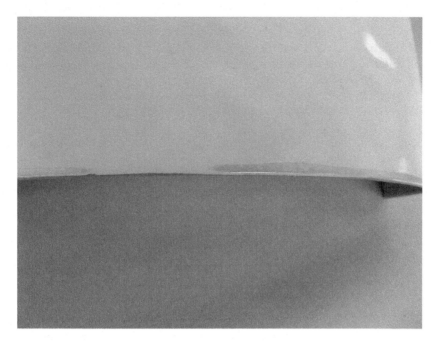

Figure 3.59 This panel was not adequately primed; as a result, of this the filler can now be seen. To rectify this the whole panel will require re-priming. Once the panel has been cleaned etch primer needs to be applied to the bare metal before the high-build can be re-applied.

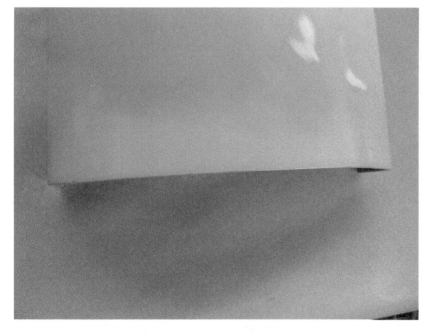

Figure 3.60 Rub-through primer has been applied to the bare steel to provide corrosion resistance before being recoated with high build to even out the filler.

rub-throughs are there on the panel? How many rub-throughs are *too* many should be left up to the experience of the technician. But as a rough guide, if the areas of rub-through primer are starting to meet and merge into one, defeat should be admitted and the panel primed properly.

3.14 FOUNDATION MATERIALS CONCLUSION

There are many different types of foundation material, and each has a very particular job to do. Some are used for adhesion, some for flexibility, and some for corrosion resistance through chemical protection or waterproofing. Other foundation materials are used to help even out panel deformation, or generally help improve the final finish by providing a very smooth and even surface. Compatibility should be considered with the material that the foundation is about to be applied over, as well as what is going to be applied over the top of the foundation material. As with all refinishing materials, foundation coats should be applied with the recommended tools or equipment, and the best place to find this is on the TDSs, which are available for virtually all refinishing materials from the suppliers, or online.

Chapter 4

Preparation of panels for top coats

The preparation of vehicle panels is arguably as important to the overall appearance of the vehicle as the final finish of the top coat. It is therefore very important to choose the correct tools, materials and equipment for the job at hand.

So what is panel preparation? Simply put, it is ensuring that a panel is ready for any material that is to be applied next. Figure 4.1 shows a vehicle that has been fully prepped and is ready for top-coat. This means making sure that there are no contaminants, loose materials, moisture, residues, or corrosion present on the surface. It is also extremely important that the panel be sanded in such a way that it will provide adequate mechanical adhesion, without any scratches showing through the final coats that are applied after the sanding. The scratch marks may be visible as either straight scratch marks or 'pig tails' from the dual action (DA) sander.

We have covered the preparation to some degree in Chapter 3. In Chapter 2, there is a statement that the finest abrasive practicable should be used when rectifying defects.

Figure 4.1 This vehicle has had all preparation done ready for top coat.

The opposite is true for the preparation of foundation and original equipment manufacturer (OEM) materials for top coats. There is a fine balance with abrasives between coarse enough to provide good mechanical adhesion and too coarse, which may result in scratch marks showing through the top coats. Conversely, too fine an abrasive will likely result in poor adhesion. It is therefore recommended to use the coarsest abrasive possible without affecting the final finish.

As a very general rule of thumb, follow the advice below:

- P800–P1000 scratches will be covered perfectly by all top coat materials.
- Clear coats are likely to show the scratches of anything coarser than P800
- Base coats are fine on surfaces prepped with P400 on a DA sander or P800 (or finer) wet-and-dry abrasives.
- The scratches from grey abrasive wool are usually covered easily by clear coats and base coats; however, on certain light metallic colour it is occasionally necessary to use the finer grade, which is gold-coloured abrasive wool.
- Direct gloss can be applied over panels flatted with P320 on a DA sander, although it is recommended to use P400 instead, or P600 wet-and-dry.
- Direct gloss can also be applied over paint prepared with red abrasive wool with no visible scratches. This grade of abrasive is too coarse for clear coats though, and it is advisable not to prepare for base coats with the red grade either.
- In all the above instances, these are the coarsest grades recommended, and often a slightly finer grade is used.

The industry standard for the preparation of panels used to be wet-and-dry abrasives and a lot of soapy water to prepare all panels. Although wet-flatting procedures are still used in some workshops, they are not used often, the main reasons being the following:

- Wet flatting means wet floors and, in turn, slip hazards.
- Fillers and primers can soak up the water and trap it against the metal, causing corrosion issues later down the line.
- The water gets caught behind any trim pieces and rubbers, and can leach out during the painting process, ruining the finish.
- While flatting down, a sludge is created, which needs to be cleaned off regularly to check the progress of the sanding.
- Constant exposure of hands to water contaminated with paint sludge can cause dermatitis.
- Clean up after wet flatting takes much longer than after dry flatting.

All these points aside, flatting panels with water can have their advantages:

- The water itself will obviously clean the panel. However, soap is also often added, which helps clean contaminants from the panel.
- Gentle sanding by hand (without a block) is less likely to cause rub-through of existing paint layers, especially along swages and panel edges.
- The slower nature of hand sanding can help the technician concentrate on the panel and spot small defects such as stone chips, small scratches, pinholes and peeling clear-coat edges.

- Wet-and-dry can be used on very small intricate areas and components when using it by hand, as the fingers of the technician can be used in tight areas.

When using wet-flatting techniques, there are a few points to remember which can really help with the quality of the sanded panel:

- Keep everything free from grit. This includes the bucket, block and abrasive paper itself, and the panel to be worked on. If a 'squeak' is heard whilst sanding, it is very likely to be a piece of grit between the paper and panel, or the corner of the abrasive paper has folded over and is scratching the panel.
- Add washing up liquid or a similar soap to the water – this helps lubricate the paper, cleans the panel of some contaminants, and keeps the abrasive clear of sludge.
- Never put abrasive down on a workbench, floor, or any horizontal surface when not in use, as it will pick up the dirt and grit that is found in a workshop's atmosphere and transfer it to the panel and flatting water. If the technician wishes to put down the abrasive, they should simply put it back into the flatting bucket.
- Get the level of water correct on the panels. Too much water and the floor around the job will become dangerously slippery, and too little water will cause the sludge to build up and hold the abrasive off the panel, thus preventing it from doing the job of flatting.
- Use the edge of a rubber flatting block as a squeegee in order to remove the wet film. Then dry the surface using a microfibre cloth or mutton cloth. This is a quick and efficient way of drying the panels and is a good opportunity to check the condition of the paint finish thoroughly.
- Remember that it is not always necessary to use a full-size rubbing block. If the job at hand doesn't require primer to be shaped and is simply being undertaken to provide a key for the next coat of paint, flatting with the palm of one's hand is perfectly acceptable.

4.1 WET-FLATTING EQUIPMENT

Part of the appeal of using wet-flatting techniques is the simplicity of the equipment involved. It is usually cheaper than dry-flatting equipment, and doesn't really need much maintenance in order to last for years. Indeed, in the case of flatting blocks, ones which have seen years of use often give a better finish due to their worn edges.

4.2 BUCKET

It is important to have a reservoir of water when wet flatting. The most obvious thing to use is a bucket (Figure 4.2). However, buckets often get used by colleagues wishing to wash cars, so they should always be thoroughly cleaned before use to prevent cross contamination. Large mixing cups provide a useful alternative. In reality, any clean receptacle that will hold water and is large enough to get one's hand in can be

Figure 4.2 The bucket is a very important part of the wet-flatting process. Note the added washing up liquid for lubrication.

used. Some workshops insist on the use of water sprayers for wet flatting in order to cut down the risk of slip hazards, but this makes it very difficult to fully saturate the abrasive paper.

4.3 FLATTING BLOCK

As has been mentioned already, a good quality rubber block (Figure 4.3) will last for years. They should always be kept clean, and if they do get damaged through carelessness (like panel beaters jamming them in panel shuts to help bend a panel for instance), they should be disposed of. The author has even seen refinishers trying to achieve a good finish using blocks with blobs of body filler and paint on them! The bottom of the block should be perfectly flat with no undulations.

Some refinishers (the author included) opt to make very small blocks from wood for use in tight areas. These little homemade blocks are especially useful for removing runs. If you do choose to make one, ensure all of the edges are sanded into radii, and all of the flat sides are sanded smooth and even.

One of the most useful 'blocks' is the technician's hand. One's hand can conform to the shape of the panel and prevent grooves from the edges of a block. This technique is of course only to be used when a panel doesn't require any kind of reshaping. It is important to remember, when flatting by hand, to use the palm of one's hand and never the fingers, as these will cause 'finger furrowing'. The only exception to this is when flatting in a very tight and intricate area – then it is acceptable to use one's fingers.

Figure 4.3 A hard rubber block designed for wet flatting will provide years of service. Keep it clean and cut or chip free.

4.4 DRYING CLOTHS

As stated earlier, the edge of the flatting block is very useful for cleaning the water film off the panel. Of course, a proper squeegee may also be used but is often unnecessary. Once the majority of the water is removed, a clean microfibre (Figure 4.4) or mutton cloth should be utilised to clean the panels down. This is preferential to using paper towels, as they are costly and do not dry the panels as thoroughly as a microfibre, and there is also a slight danger of cheap paper towel scratching the paint.

4.5 WET-AND-DRY ABRASIVES

These may be bought in different formats, but the most common are A4, A5 and as a roll (as can be seen in Figure 4.5). When technicians require a piece of wet-and-dry, they usually tear off the required amount, but some refinishers argue that the abrasive paper should always be cut with scissors rather than torn, as the resultant rough edges can scratch the paint surface during use. Regarding tearing or cutting abrasives, the author considers it to be an unnecessary worry to have to use scissors. The paper should be folded neatly and firmly, then torn carefully. The edges of the paper should be folded up and kept away from the panel, and as has already been described, if a squeak is heard, it may well be the edge of the paper scratching the panel.

Figure 4.4 A wet microfibre provides fast cleanup of wet-flatting sludge. Wring it out and it will dry the panel fast too.

Figure 4.5 Wet-and-dry abrasives are available in many grades. They are usually supplied in either A5 or A4, then torn down to size.

4.6 DRY FLATTING

As previously mentioned, wet flatting was the industry standard; now, however, dry-flatting techniques are the norm. The reasons are many but include the following:

- No water on the floor causing slip hazards.
- Bare metal can be safely sanded without concerns of promoting corrosion.
- No flatting sludge means a far faster cleanup.
- No water is stuck behind trim, gaskets and rubbers.
- Powered sanders can be used without concerns of splashing dirty water around the workshop.

Dry flatting is not without its issues though. The dust created is hazardous to health, with the respiratory system most under threat. This is why it is advisable to *always* wear a dust mask when dry flatting, even if there is adequate extraction. On the subject of extraction, it should be stressed that efficient extraction should always be used when practical – basically that means during all procedures except when using foam-baked abrasives by hand. This creates another negative point for dry flatting – the need for extraction, with either a mobile unit or a central extraction unit which is piped around the workshop. Either type is expensive and creates the trip hazard of trailing hoses, as can be seen in Figure 4.6. This is magnified when using powered sanders with either pneumatic hoses or electric cables trailing.

Dry-flatting abrasives are more expensive than wet-and-dry abrasives. This is because dry abrasives require cutting to shape for sanding blocks or DA sanders (Figure 4.7).

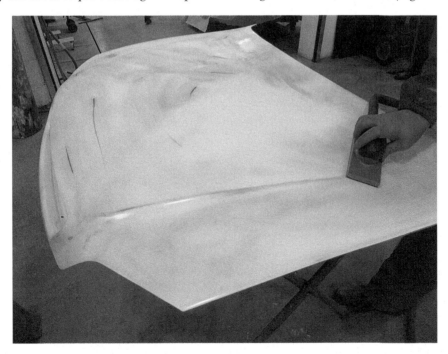

Figure 4.6 This bonnet is being dry flatted with an extracted block in order to level the surface prior to finishing with a DA.

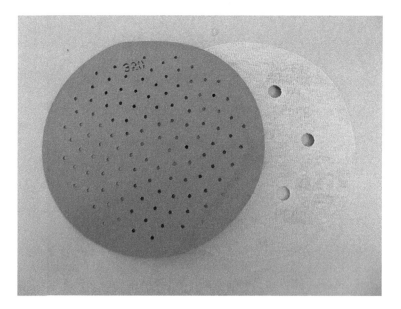

Figure 4.7 DA discs are attached with hook and loop. Notice the fluffy texture to the back of the bottom disc in this picture.

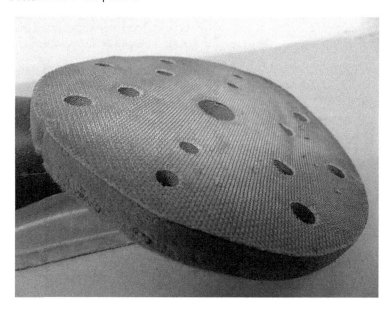

Figure 4.8 This interface pad clearly shows the hooks that the DA disc attaches to. The other side of this pad has the fluffy texture to grip the hooks on the DA backing pad.

Dry-flatting abrasives also require an attachment mechanism, usually a hook-and-loop system as in Figures 4.7 and 4.8, but occasionally adhesive is used. A perforation system for extraction purposes is also required, and the abrasives are backed with a cloth material rather than a paper. This makes dry-flatting abrasives far more robust than wet-and-dry abrasives, and this is especially important for powered sanders.

4.7 DRY-FLATTING TECHNIQUES

Dry-flatting techniques are varied and include the use of DA sanders of various sizes such as the one pictured in Figure 4.9, flat bed sanders, sanding blocks and hand sanding.

DA sanders are often referred to as random orbit sanders due to their double rotating pattern. It is this pattern that makes them so suitable for both preparation and final finishing work. It has been said that a DA sander leaves a finish equivalent to blocking with abrasive paper twice as fine. Although there is no evidence to corroborate this, it is true that a panel flatted with P500 on a DA sander will have a finer, less scratched surface than that of a panel flatted by hand and P800 foam back abrasive.

DA sanders are commonly used with backing pads of 80, 125 and 150 mm, with available abrasive grits varying from P40 to P4000. Figure 4.10 shows some examples of the size variations and different looks of the abrasives. As a general rule of thumb, grades P40–P180 are used for stripping previous materials back to the bare substrate or flatting filler work. Grades P240–P500 are used for flatting primers back, ready for top coats. Grades P800–P1000 are used to prepare previously painted surfaces such as OEM finishes, ready for blending and clear-coat application. Finally, P1200 and finer are used for flatting panels ready for polishing.

There are many factors to be taken into consideration when deciding on the size of backing pad, grade of disc, and whether to use an interface pad. The decision to use an interface pad or not will rely on the many variations of finish of the surface, the type of repair and the shape of the panel. Interface pads (commonly referred to as soft pads) are simply a circle of soft foam having the same diameter as the disc and backing pad, with matching extraction holes in place. On one side is the hook fastening to attach to the DA disc, and on the other side is a loop fastening to attach to the backing pad of the DA sander. Interface pads offer conformity, which means that when performing final prepping for top coats, the panel contours are less likely to

Figure 4.9 The DA is frequently used for dry flatting.

Figure 4.10 Mini DAs are handy for restricted areas. The full-size one is in this picture for scale. The mini DA in this picture is of limited use due to a lack of extraction.

be misshapen by removing too much material from one area. Interface pads are also very useful when feathering out the edge of a repair, such as around a key scratch, stone chip, or, more important, the feathering of the edge of an area of patch primer.

When using a DA sander to prepare a panel for top coat, it is important to consider the shape of the panel. This is true whether you are using an interface pad or not. It is also important to consider whether there is primer on the panel, whether the panel is to be painted edge-to-edge or blended, and what type of top coat material is to be used. The following table should help with the choice of grit size of the abrasive disc. Remember – sometimes the coating being prepared will require a coarser grit in order to provide a mechanical adhesion, and sometimes the colour may dictate a finer grade, to prevent visible scratch marks, commonly referred to as 'pig tails' due to their curly nature (Figure 4.11). As a general rule of thumb, the coarser the grit, the better, as the larger scratches provide the best possible mechanical adhesion there are some examples of different abrasive grades pictured in Figure 4.12. Obviously, a primer patch should be blocked first, and the blocking marks will require removal with a DA disc, when a P500 is usually used, as primer will need to be coated with colour.

Type of top coat material to be applied	Grit size of DA disc
Clear coat (the part of a panel having no base coat applied)	P800–P1200
Base coat (water based or solvent based)	P400–P800
Direct gloss	P320–P800

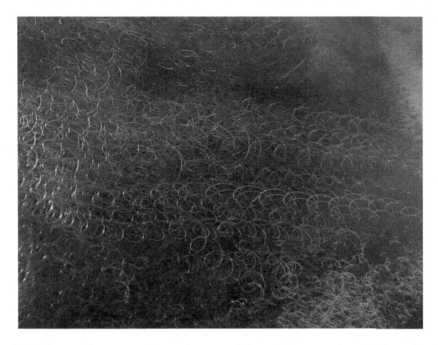

Figure 4.11 DAs provide an unusual scratch pattern. The orbit can go as low as 1.5 mm, but these scratches known as pig tales came from a DA with a scratch of around 5 mm.

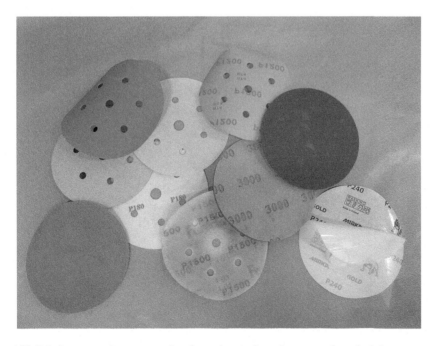

Figure 4.12 DA discs come in many grades. Some have a foam layer to soften their impact or retain water for wet flatting.

4.8 A FEW RECOMMENDATIONS FOR DA USE

- Keep as much of the abrasives in contact with the panel as possible at all times. If this means using an interface pad, then use one. This is especially important when feathering out many layers of paint like those pictured in Figure 4.13.
- Make long passes across the entire panel or primer area. This helps prevent ripples in the final finish. A DA should never be used to dig into a small area at the final preparation stage.
- It is fine to use a slightly coarser grade of disc to remove orange peel before finishing with the correct grade, but it is recommended to drop only one grit size.
- A DA should never be used to shape filler or primer. That should be done with a block or by hand.
- A DA doesn't always need to be at full speed – it is sometimes an advantage to turn the speed down and take your time.

4.9 DRY-FLATTING BLOCKS

As previously mentioned, when shaping panels, a DA is of little use, because it is both round and spinning. The most precise way to shape primer at the preparation for top coat stage is with the use of an extracted block. Of course flatting blocks are used to shape filler too as pictured in Figure 4.14. It is strongly advised that any dry flatting with a block be done while using adequate extraction.

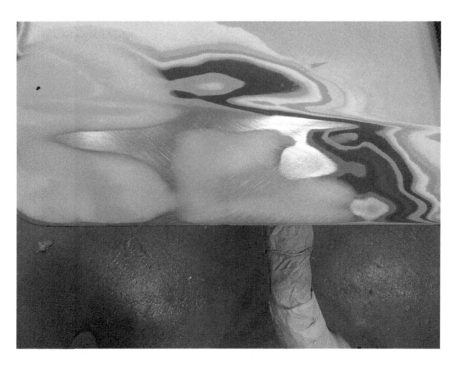

Figure 4.13 This panel has been well feathered which is no small task with so many layers of paint. The DA is very useful for this.

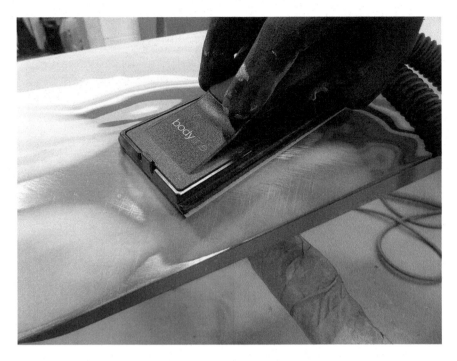

Figure 4.14 Dry-flatting blocks are largely used for shaping filler and smoothing primer.

Extraction blocks are available in many lengths, typically 70 mm wide, while standard lengths are 125, 198 and 420 mm. The longer ones can often be shaped through their length with the use of an adjuster, and the shorter ones are available in concave and convex shapes, as well as the more usual flat type.

When using a flatting block, it is important to remember where the edges of the block are, as it has been observed on many occasions, that while trying to shape an intricate contour with the flat of the block, the edge is digging into another part of the panel. When flatting primer to get it ready for top coat, it is advisable to plan how the passes are going to be made. At this point, applying a guide coat is strongly recommended for lighter shade primers like the bonnet pictured in Figure 4.15, while for black and dark grey primers, it is usual not to bother with a guide coat, as it is much easier to see the imperfections in these darker shades.

Guide coats are available as a dry powder applied by sponge and a dry powder applied by aerosol; although not officially a guide coat, satin black paint is often used from an aerosol. It is usual to start sanding lightly across the panel in order to get a feel for the shape and highlight any slight high or low spots (Figure 4.16).

Then sand the primer further in a diagonal pattern from the top to the bottom (Figure 4.17) – it doesn't really matter if one starts from the back of the panel working forward or from the front working backward.

However, what *is* important is that a second pass is performed diagonally in the opposite direction to the first as in Figure 4.18.

This pass should be the one that finishes the shaping procedure. If it requires further sanding, repeat these diagonal passes. It is quite OK to then lightly sand in the

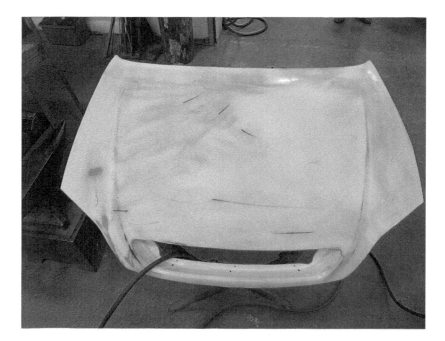

Figure 4.15 Dry guide coat is applied to this white primer. The powder fills the previous scratches uniformly. Note the trailing extraction hose and airline associated with dry flatting.

Figure 4.16 Initial sanding should be done lightly across the panel in order to gain a feel for the shape.

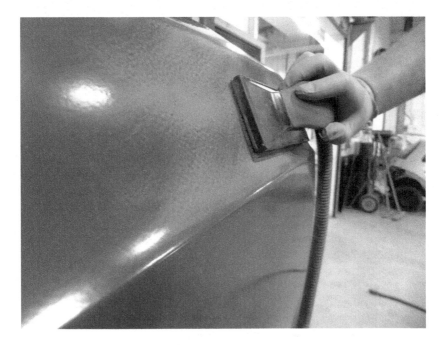

Figure 4.17 First sand diagonally in one direction.

Figure 4.18 The second pass should be diagonal in the opposite direction.

front-to-back direction again. The reason to sand diagonally in two different directions is to prevent panel undulations.

Using a block up and down will give a very wobbly final shape along the panel length, whereas using a block purely front to back can lose the curves associated with vehicle bodies, and instead create a series of straight edges (Figure 4.19) – commonly described as being like a fifty-pence piece.

The repeat pattern of diagonal flatting is known as 'cross-hatching' the panel.

It is recommended to perform primer shaping with P320, which is usually coarse enough to make rapid progress, and yet fine enough not to leave very deep sanding scratches. However, even P320 scratches show through base coat, so always reapply a guide coat. At this stage, the author highly recommends a dry dust guide coat from a sponge applicator, as it highlights the finer scratches much better than either aerosol types. Next, use P500 on a DA with an interface pad to remove those P320 scratches (Figure 4.20). The interface pad is necessary to prevent accidentally reshaping the panel with the DA, as the soft pad makes the DA disc conform to the panel's shape, while still removing the P320 scratches.

In final preparation tasks, a sanding block may be utilised to flat small areas of stopper that have been used to fill stone chips (Figure 4.21). When used in this way, the block should be held almost 'hovering' over the panel, so that the stopper is taken down with the lightest touch, and this should prevent breaking through the surrounding paint films.

Figure 4.19 Be careful not to flatten off the curve of panels by sanding back and forth on curves.

Figure 4.20 Although this panel is not blocked, the nice consistent scratch offered by P500 on a DA is clearly evident.

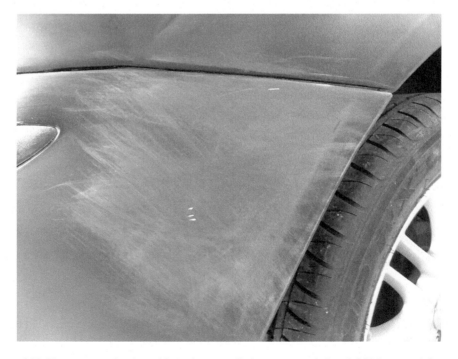

Figure 4.21 The stopper clearly visible in these small chips was sanded with P800 wet and dry.

4.10 A FEW RECOMMENDATIONS FOR USE OF DRY-FLATTING BLOCKS

- Choose the block wisely. Do you have the correct length block? Should you be using a flat block – which is most usual, or a concave or convex block?
- Sand *with* the shape of the panel. Wheel arches are possibly the most complex shape to flat – don't be scared to follow the curve.
- Use the cross-hatching technique, as it prevents wobbly panels and flat spots where there should be curves.
- Use a guide coat. Experiment and find which type works best for you. Dry powder guide coat can provide a type of lubrication which helps reduce the effort required in flatting.
- Always use extraction.
- Let the block do the work – don't push down on it too hard, as this will cause gouges and speed up the fatigue of the technician.
- Don't be afraid to move the abrasive strip on the block (Figure 4.22). For instance, when flatting inside a concave area, it is sometimes handy to have the strip slightly up past the edge of the block as can be seen in Figure 4.23.
- It is very unusual to top coat directly over blocked panels. It is typically necessary to finish with a DA or foam-backed abrasives by hand.

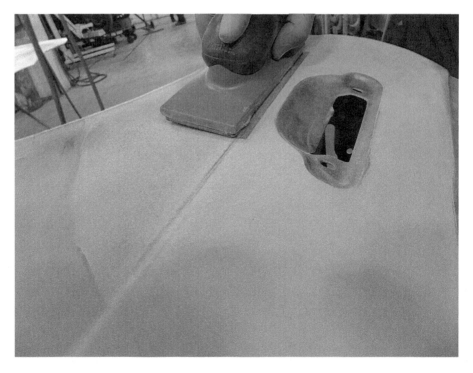

Figure 4.22 Notice how the abrasive strip has been positioned to one side of the block; this helps sanding in the swage.

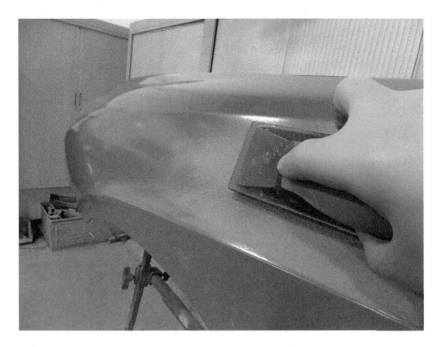

Figure 4.23 On this Bentley wing, there is a subtle concave curve which is sanded diagonally. The abrasive strip is off set to work inside the radius.

4.11 ORBITAL PALM SANDERS

Orbital palm sanders are available in a variety of sizes. They are powered by either electricity or compressed air. It should be considered that they are not always designed purely with the automotive environment in mind; therefore, it is important to choose wisely if you decide to purchase one.

The most important factors are as follows: the availability of abrasives in the size of the bed, what method is used to attach the abrasive strips, the level of extraction and the size of the orbit (around 3 mm gives an excellent finish for top coats). There are many manufacturers producing these for 70 mm×198 mm sheets, which is also one of the most popular sizes of hand-flatting block. This means that the abrasive strips are interchangeable.

Orbital palm sanders are occasionally known as powered blocks, as they are used in exactly the same way as a dry-flatting block, and can be utilised in the shaping of both fillers and high-build primers. With this in mind, the same rules apply for using these, as for dry sanding blocks. Indeed, orbital palm sanders should be considered suitable to do the same job, but with different pros and cons.

- Pros:
  - They do not leave the straight scratch mark of a hand block.
  - They do some of the work for you by cutting in an orbital motion, and coupled with the back-and-forth motion of a hand-powered block. Further, to speeding up sanding times, the extra abrasion also cuts down on the fatigue of the technician.

- Cons:
    - Technicians often complain of having less control, as the block is taking down material too quickly.
    - Some refinishers have suggested that they are not as accurate for shaping swages and around wheel arches. This is due to the orbital movement of the block.
    - They are heavier than a hand-sanding block.
    - They often have a brittle composite backing bed that is easily broken if dropped.

Orbital palm sanders definitely have their place in the preparation of under coats for top coats. However, due to being halfway between a sanding block and a DA, they are often overlooked in favour of the precision of a sanding block followed by the smooth finish of a DA sander.

4.12 ABRASIVE POLYMER WOOL

So far in this chapter, we have looked at the tools and abrasives used to remove very thin layers of material in order to shape, even out and smooth off foundation coats and previous coatings. However, in the preparation of panels for top coats, blend panels need to be discussed.

A blend panel is one that will have base coat applied to one end followed by clear coat, whereas at the other end, there will be clear coat only over previous paintwork. The preparation is often undertaken by the use of fine-grade abrasive papers on a DA, but this can also be done by hand. There is another very useful alternative that removes even less of the material, and simply provides the minute scratches required for mechanical adhesion. For this purpose, abrasive polymer wool is ideal.

At this point, there will be many readers thinking, Surely they mean Scotch-Brite?' Well, yes I do, however that is a trade name of 3M, and only their product can be called this, even though every workshop and supplier refers to it by its common name. With that in mind, an 'umbrella' term for all manufacturers of woven hand pads is abrasive polymer wool. The wool used is often polypropylene, and the impregnated abrasives can be aluminium or titanium oxide among others.

Although these hand pads are made by many manufacturers, they have a clearly defined grading system specified through colour, as shown in the following table. It should be remembered that all technicians and workshops have their own definitions

Colour of abrasive pad	Use in preparation of panels for top coats
Gold	Flatting previous top coats ready for application of fresh clear coat, provides a very fine scratch for use on softer clear coats and colours that show scratches easily (Figure 4.24)
Grey	Flatting previous top coats ready for application of fresh clear coat, and lighter metallic such as silver (Figure 4.25)
Purple	Not usually used under clear coat, can be used under most metallic colours
Red	Suitable under direct gloss colours, and very occasionally base coats, also used to prepare previous coatings for primer coats (Figure 4.26)

Figure 4.24 Copper or gold polymer wool is used when clear coating over difficult OEM colours or often in fade-outs.

Figure 4.25 Grey polymer wool offers a finer scratch it is used where clear coat is to be applied.

Figure 4.26 Red polymer wool is the coarsest used for most jobs in the refinishing industry.

of where each of these can be used; however, the table shows the generic use which most technicians can agree on.

Of the four grades shown, grey and red are by far the most popular as they have little overlap of use. These abrasive hand pads are often used dry and cut very well in use. However, many technicians like to use water with the pads to provide lubrication and cut down on dust. Further to this, a flatting paste (Figure 4.27) can be used on the pad (Figure 4.28) to provide a more uniform cut (Figure 4.29) from the abrasive wool, and this is particularly useful on well-cured, hard top coats. Flatting paste also has the useful attribute of cleaning contaminants off the panel at the same time. This is most useful for panels that have ingrained dirt on the surface.

4.13 PREPARING FOR A BLOW-IN

When painting certain panels, it is more cost and time effective to perform a 'blow-in', more correctly referred to as a fade-out, and this will be fully covered in Chapter 7. A brief description is applying the gloss coats (clear coat and direct gloss) to a panel, without painting to the panel edges. If this is done correctly, it will produce a seamless repair – however – the key is good preparation. Fine-grade abrasives should be used, and every refinisher has their preferred method, but usually, the part of the panel not receiving top coat should remain shiny and as such is often polished first. The area around the colour should be prepped with grey abrasive wool or abrasive paper P1500–P2000. This can be done by hand or with the use of a random orbit sander. The panel should then be thoroughly cleaned, as with all other preparation procedures.

Figure 4.27 Flatting paste is applied to wet polymer wool. It helps increase the matting effect of the wool.

Figure 4.28 The wet polymer wool is rubbed with the flat of the refinishers' palm across the panel to be blended.

Figure 4.29 The difference in reflection can clearly be seen around the edge compared to the centre of this picture. The scratch pattern is further enhanced by the flatting paste.

Figure 4.30 When panels have a matt appearance due to being sanded, it is difficult to see reflections. This focus has been wetted with a solvent panel wipe to provide a gloss in order to check straightness.

4.14 FINISHING PANELS FOR TOP COAT

Once the foundation materials and blend areas are flatted, smoothed and keyed, using the techniques and materials discussed in this chapter, it is very important to check that all panels are completely ready. It is easy to miss a small area such as an indicator or door handle recess, or sometimes a stone chip or light scratch which isn't included in the initial repair. As the vehicle will need to be meticulously clean for paint application anyway, it is best practice to clean all panels with a panel wipe using the wipe-on-wipe-off technique.

This means using a cloth wet with panel wipe to clean the panel and a dry one immediately after to wipe off the residue. When the panel is wet with panel wipe, it will appear shiny if one looks down the side contours, and while it is shiny, it is possible to see if the correct shape and profile have been obtained (Figure 4.30). This process can be used after initial sanding, as well as at the end of the preparation work. If a wobble or incorrect profile is spotted prior to finish sanding, it can be rectified more quickly and easily. Whether you are preparing panels or an entire vehicle for top coats, it should be remembered that paint will *magnify* any sanding marks or incorrect contours – not hide them. It is therefore more material and labour efficient to ensure *all* preparation work is perfect before masking the vehicle.

Chapter 5

Masking of vehicle and components

When masking vehicles, it should be remembered that there are two reasons for doing so. The first and obvious reason is for preventing paint and overspray from getting onto panels and parts that are not part of the repair. The vehicle featured in Figure 5.1 required nearly the entire body repainting however plenty of masking was still used to protect the components which remained. The second is for protecting the fresh paint film from contaminants found on the rest of the vehicle. Regarding contaminants, it should be noted that every effort should be taken to remove them from the entire vehicle prior to initial preparation work.

When panels are prepared for paint, there is often a need to weigh up the options of whether to mask the component or remove it from the vehicle. In an ideal world,

Figure 5.1 Masking is a vital step in vehicle refinishing, enabling components to be sprayed and others to remain unaffected.

all components would be removed from panels prior to preparation, but sometimes the labour time involved in removing a component makes it prohibitively costly to do so. On other occasions, the component could be extremely expensive and delicate, and the risk of damaging it during removal could result in loss of profit from the job. Figure 5.2 shows a heavily masked bumper, it was decided to be too time inefficient to remove the front bumper, lights and number plate plinth, so accurate masking was used instead.

Masking requires precision and patience in order to provide a seamless repair. If any masking material encroaches onto the panel which requires painting, there will be marks left in the fresh paint film – this is unacceptable. Conversely, if there are any parts of the component visible prior to painting, these will get paint on them, ruining the component and subsequently requiring extra labour time to clean it up.

There are many different types of masking material available to the refinishing technician, and each individual will have their preferred materials and preferred methods and techniques of using them, so much so that it is virtually impossible to list all of them.

5.1 TYPES OF MASKING MATERIALS

5.1.1 Masking tape

Masking tape (Figure 5.3) is usually made from a crêpe paper-type backing with an adhesive of acrylic or solvent rubber. It is used for everything in the workshop from masking vehicles, to labelling parts, and even makeshift plasters for small cuts! It is available in many sizes, ranging from 18 to 100 mm, with the most popular sizes being 19, 24, 36 and 48 mm. Masking tape may be used in conjunction with other materials such as 'stat' sheet or masking paper, or on its own on smaller components such as door handles or lights.

Figure 5.2 Masking can save time when painting components that are very complicated to remove.

Figure 5.3 Most masking tape is paper based and comes in a variety of sizes.

5.2 FINE LINE TAPE

Occasionally known as 'low tack' tape, fine line is made from plastic. Different manufacturers use different plastics these appear as different colours as in Figure 5.4; some are thicker and more elastic than others; the standard sizes are 3, 6 and 9 mm. The physical properties of the plastic tape make it much more flexible than paper-type masking tape. This is very useful when masking complicated shapes, whether it is around rubbers and body gaskets, two-tone paint jobs, or custom graphics. Another useful property of fine line tapes is their ability to prevent paint bleed. This is where paint creeps under the edge of masking tape leaving a blurred line.

When working with fine line tape, technicians will often remove the tape shortly after the final flash off, before the booth goes on bake. Removing the tape while the paint film is still wet helps prevent bridging between the fresh paint and the masking tape. It will also allow the paint edge to flow slightly flatter than if the tape is left in place. In this instance, the plastic nature of the tape again comes into its own as it can cut through the cured paint film easier than paper masking tape. Removing tape from wet paint film is not for everybody though, and some refinishers would rather leave it in situ rather than risk dropping tape in the wet paint.

Fine line tape is not used to replace paper masking tape completely. It is far more expensive than the paper type and obviously much narrower, and if it was manufactured in the larger sizes, it would be extremely difficult to tear, even in the smaller widths. It is therefore most often used in conjunction with conventional masking tape around glass, gaskets and door handles.

Figure 5.4 Fine-line tapes are made by several manufacturers. These manufacturers often have their own colours; fine line tape is sized in 3 mm increments.

5.3 MASKING FOAM

As the name suggests, masking foam is made from cellular foam and is generally around 12 mm in diameter. Most manufacturers produce their masking foam with a line of adhesive around 5–6 mm wide down one edge. This is dispensed from a box which usually contains many rolls side by side. The one in Figure 5.5 has one roll left in, in order to better illustrate the size and length of each roll.

Some manufacturers produce foam with no adhesive whose primary function is to block the gap between the trailing edge of the front wing and the leading edge of the door. Adhesive-type masking foam can be used to provide a soft edge for patch priming or clear coat blow-ins, as an easier alternative to folding the edge of standard tape over. It is most often used along the edge of a shut line and is designed to block the gap where two adjacent panels are being painted together. This prevents overspray from entering the interior of the panel shut and gives a slight fade-out of top coats on the actual transition between the outer and inner panels.

When applying the foam tape, great care should be taken to place it in exactly the right place. Too far into the shut, or sticking out of the panel gap, will create a very unsightly hard tape line on the edge of the panel.

5.4 MASKING PAPER

Masking paper is a lightweight brown paper. At one time, it was used to mask out vehicles in their entirety; however, that function is now more usually performed by

Figure 5.5 Masking foam is manufactured from cellular foam; it will fill gaps between panels. There is mild adhesive down one edge. The round cross section offers the ability to create soft edges for patch priming.

static masking sheet. Nowadays, masking paper is most often used on windows, mirrors, lights, grills and wheels. Masking paper is available in many sizes from 150 mm (6 in.) to 1,200 mm (48 in.). The most popular sizes are 450 mm (18 in.) and 900 mm (36 in.), as shown in Figure 5.6.

When inspecting masking paper, it has a very obvious outside and inside – one side is shiny and the other side is matt. Masking paper should always be used the shiny side out, the reasons being twofold:

1. The shiny side is less absorbent and helps prevent wet paint soaking through to the protected panel.
2. The matt side has very small, loose fibres which can come loose when agitated by the force of the spray gun. They can then land in the wet paint film, contributing to the dirt nibs.

Masking paper is generally used in conjunction with masking tape to hold the edges in place. It should always be applied neatly, any pleats or creases must be taped down to prevent dust pockets, and it should be wrapped as closely to or around the component as possible. Figure 5.7 shows a vehicle that as been masked using precise folded masking paper.

5.5 STATIC MASKING SHEET

Commonly referred to as 'stat' sheet, this very thin plastic is designed to both cling to the vehicle and attract overspray and dust through a static charge. It is most often

Figure 5.6 Masking trolleys provide a means of storage for bulky materials and a convenient way to dispense them.

Figure 5.7 Masking paper was chosen in this instance as most of the car was being repainted, so it made more economic sense to use several sections of paper rather than a piece of static masking sheet.

Figure 5.8 Static masking sheet offers quick convenient masking of large areas of a vehicle; it is more expensive than paper but much faster to use.

supplied in 3 and 4 metre widths, and as such is folded over repeatedly on the roll, in order to make it more manageable to store and use. The outer edges are to be found in the middle of the underside. Once these are pulled out, it will become evident that the sheet is still doubled and will require unfolding from the middle a second time. At this point, it is best practice to pull the sheet tightly over the vehicle and secure all four corners firmly with an overhand knot. This not only prevents the sheet flapping around and ballooning up in the cross breeze of drying cycles but also makes it much easier to cut the sheet around the panels and components to be painted.

Stat sheet is printed with instructions for use, with one side of it facing outward. This can be seen in the Figure 5.8. When considering the economics of material use, stat sheet is an excellent choice for most jobs and only becomes less economic when over half of the vehicle is to be painted. Stat sheet is usually 3–4 m wide and therefore provides more than enough cover for two sides and the roof.

5.6 LIFTING TAPE

Lifting tape is a very specialist masking material designed for raising the edges of window rubbers and body gaskets. It replaces the more traditional method of forcing string or wire under the edge of the rubber, and then adding tape over the rubber. By lifting the rubber, it is possible to fade out the edge of the top coat under the seal, thereby eliminating the hard edge associated with a buildup of paint one inevitably gets around any deviation on a panel.

Pictured in Figures 5.9 and 5.10 are two types of lifting tape – the one in the small bucket works on a similar principle to the old technique of forcing string under the edge of the rubber seal. This material has been specially designed to lift the seal cleanly away from the panel before it is masked out with tape. This specialist material has no fibre to it, doesn't squash down like string or wire, and can be reused time and time again. The other type, shown on the right, is far more widely used. It is made of 50 mm (2 in.) masking tape, with two strips of stiff plastic stuck down one edge. The first of these strips is forced under the rubber and is used to lift it, and then the tape is stuck to the window in order to keep the seal raised from the panel. Many manufacturers of masking materials produce the same product with only a single strip of plastic down the edge. This is used in exactly the same way.

5.7 WHEEL MASKS

During vehicle masking, many different techniques are used to mask wheels. Some people prefer to mask off the back of the wheel arch with wide tape, and then apply masking paper to this. The disadvantage of this is that the tape will not stick to the road grime invariably found behind wheel arches, which, of course, should be cleaned prior to preparatory work, but they are often either missed or inadequately cleaned. This then results in the tape failing and overspray entering the wheel well. Figure 5.11 shows an example of wheel arch masking which will not peel off to expose the inner arches and suspension components.

Figure 5.9 This is an unusual form of lifting tape, which is much like the older method of forcing string under the rubber and then using tape to cover it. This plastic strip is less likely to cause dirt nibs though.

(a)

(b)

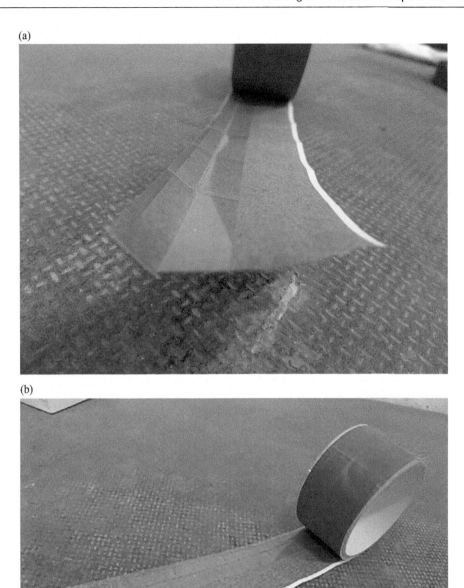

Figure 5.10 (a) Lifting tape has a stiff plastic edge which can be used under the rubber; the tape is then pulled up over the rubber to protect it. (b) The lifting tape comes on a role like most masking tapes.

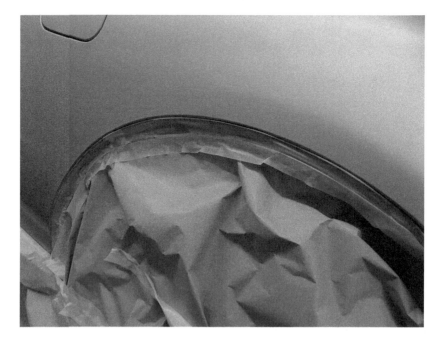

Figure 5.11 Although dedicated wheel masks are available, it is more usual to use tape and paper. This vehicle conveniently had a rubber trim on the lip of the arch which could be taped to directly.

Due to the issue of the tape not sticking properly under the arches, many technicians will opt to take a large sheet of masking paper and wrap the wheel by pushing the paper tight around and trapping it behind the wheel. This will obviously allow some overspray into the wheel well, and many refinishers have their own techniques to rectify this.

Along similar lines to this technique is the use of dedicated wheel masks. These are sometimes a solid item – like the spare wheel cover of an off-road vehicle – but more often, they are made of polypropylene (the same material as many disposable spray overalls). These types have an elasticated aperture that holds the mask tightly over the wheel. Although they can be reused, it is not recommended to do so too often, as they can release dried paint onto the fresh paint when agitated by the force of the spray gun.

5.8 PRETAPED MASKING FILM

This is made from the same material as static masking sheet. It is available in a variety of sizes, usually much smaller than full-size stat sheet. As with stat sheet, it is folded on itself many times and can be opened up to cover a larger area such as a windscreen. It is used for many jobs, including masking under arches, covering windows and inside panel shuts. The constant tape down one edge, can be seen in Figure 5.12 it offers excellent uninterrupted sealing of masking, but it does make it fairly tricky to use, as the plastic sheet often gets stuck to the tape, so it should be applied carefully.

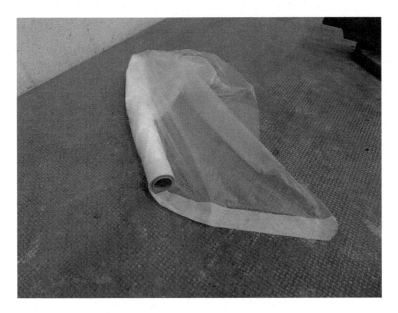

Figure 5.12 Masking film is available in smaller sizes with tape applied down one edge. This is very quick to use and thoroughly seals the components from overspray.

5.9 LIQUID MASKING

It is worth mentioning that liquid masking exists on the market. This has many advantages, one being how quickly a car can be covered by spraying the mixture over the entire vehicle, and then washing off the panel(s) that require repair. The masking can then stay on through the entire repair and paint process. When the repairs are completed, the masking is simply washed off – along with all the overspray and filler dust. It does have several disadvantages though, including its high cost and the fact that panel gaps near the repair will still need sealing or masking. Due to these reasons, it is not often used.

5.10 MASKING TECHNIQUES

As with so many aspects of the refinishing trade, there are numerous different techniques involved to reach the same end point. Refinishers all have their own favourite techniques which work best for them. This doesn't mean that their way is the only way; therefore, understanding the pros and cons of different techniques makes it possible to choose the best one for each job at hand. This can depend on whether it is the easiest, cheapest, or sometimes the one technique that results in the least finishing afterward.

The art to proper masking is to not show where the vehicle has been repaired, so it is usual to mask the edge of panels and trim in such a way that the top coat fades itself out, around the radius of the edge of panel. In the situations where a component such as a handle, light, or sealer edges contacts the panel, it is necessary to use a hard masking technique. As with all masking, it is vital that this is extremely accurate.

5.11 HARD EDGE MASKING

This can be used for two-tone paint jobs, but more often for masking components that cannot be removed for one reason or another. Depending on the proximity of the component to the panel, it may be necessary to mask adjacent to the edge with the standard masking tape and then right up to the join with fine line tape. This should then be removed, after flash-off, from the top coat to allow the edge to settle, flow and relax.

As has already been stated, it is vital to mask accurately. If the tape allows some of the components to show, this part of the job will be ruined; if the tape lies on the surface of the panel, the paint job will be impaired.

In Figure 5.13 the image shows the door handle is masked using the standard 20 and 36 mm masking tape. There is a gap of approximately 1 mm showing the rubber gasket – this will be covered with the fine line tape.

Figure 5.14 now shows that the gasket has now been covered with 3 mm fine line. Notice that the top line of tape is not a perfectly straight line. This is because it follows the gasket accurately. Once the door is painted and allowed to flash-off, the blue tape will be removed prior to baking, in order to allow the paint film to flow around the gasket. At this point, it would be prudent to use a wooden pick or plastic spreader to clean around the gasket properly to remove the ingrained dirt.

Figure 5.15 shows the window scraper trim which is masked with 20 mm tape. There is a clear gap under the edge of the trim, which makes it a simple task to wrap the tape slightly under the edge. A combination of these and other masking techniques were then used to result in the fully masked door shown in Figure 5.16.

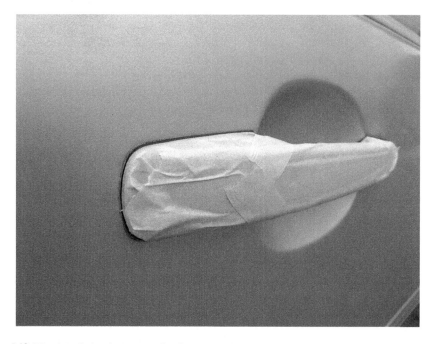

Figure 5.13 This handle has been completely wrapped in tape. Note that there is a gap of about 1 mm between the handle and the panel; this is to prevent bridging.

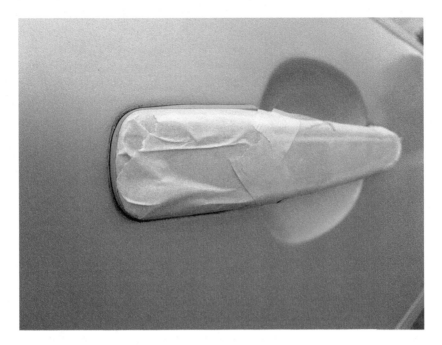

Figure 5.14 This 1 mm gap is then covered with 3 mm fine line. This offers an accurate edge up to the panel and can be removed after the clear coat has flashed off to stop bridging and flash lines.

Figure 5.15 This trim has a large gap between it and the panel; the paper masking tape can be tucked into the gap with the use of a filler spreader.

Figure 5.16 The panel is fully masked prior to the vehicle going in the booth and covered with static sheet.

5.12 BACK-MASKING

This technique is used chiefly for masking an aperture resulting from a component being removed, such as a window, handle, or light.

As well as masking apertures, Figure 5.17 shows how back-masking is often used on the edge of a panel. Its advantages are that it is very quick to do, and it seals the aperture completely to prevent the ingress of over spray, while still allowing the top coat to fade around the radius of the edge of the panel.

Figure 5.18 shows the tape being applied to the back of the door edge. This technique is called back-masking. As well as sealing the aperture, it allows the top coat to fade out around the edge of the panel. It also provides the useful property of having the adhesive side facing out, so that the stat sheet or paper can be stuck directly to it.

5.13 ROLL AND SOFT-EDGE MASKING

As stated previously, soft-edge masking may be performed with the use of a foam tape, a rolled-back tape, or sometimes a folded tape as shown in the image shown in Figure 5.19. This type of folded tape is easily made by folding the edge over, and then pulling the tape across one's knee. This tape can then be used on swage lines when a panel is to be partially painted or as a soft edge for patch priming.

In Figure 5.20, the tape is being applied to the middle of a panel in order to patch prime over a number of small marks. The use of folded tape allows any overspray to drift up to the masking and not create a hard edge, as the overspray has the chance to drift underneath the edge, as can be seen in Figure 5.21 below.

Figure 5.17 Here the aperture from the rear light being removed and the boot being raised can be protected by back-masking with 50 mm tape.

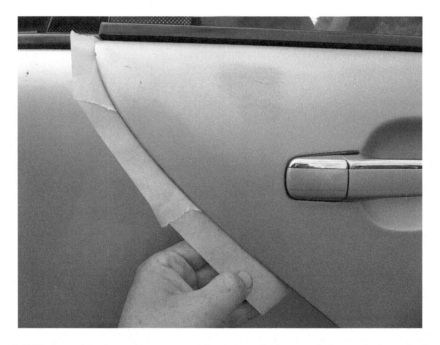

Figure 5.18 Back-masking is used to protect adjacent panels; it can only be used on panels that can be opened to provide access to the rear.

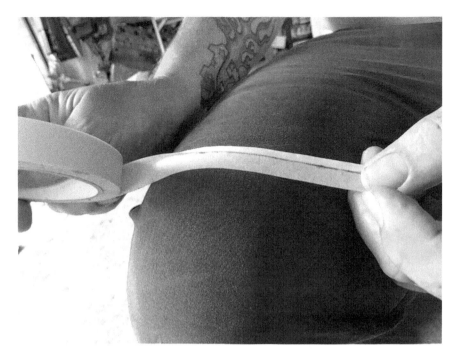

Figure 5.19 There are many techniques that can be used to create flip-edged masking. The quickest and easiest is to pull it across one's knee.

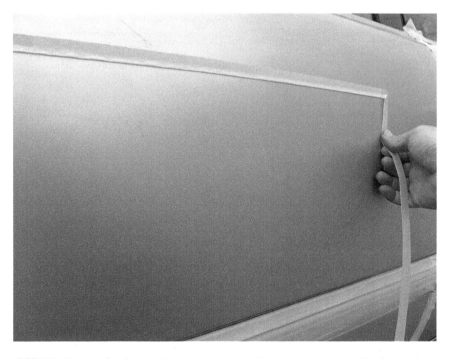

Figure 5.20 The flip or soft-edge masking tape can be used on swages or to outline for patch priming.

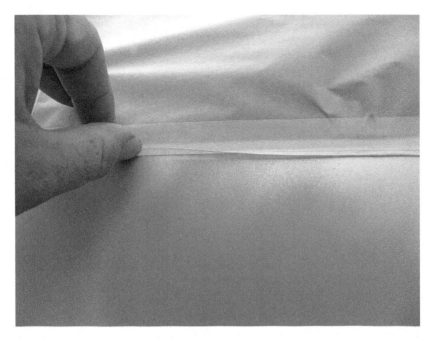

Figure 5.21 **Note how the flip edge of the tape provides a soft graduated edge to the patch primer.**

5.14 ECONOMIC USE OF MATERIALS

When considering the economics of material use in masking, it should be remembered that sometimes it is cheaper to cut the labour time and costs by using a material that is quick to use but expensive to buy; a case in point – static masking sheet. This is far more expensive than brown paper, but it takes an experienced refinisher 1 or 2 min to sheet out an entire car with it, compared to 15–20 min, to properly seal the many sheets of brown paper required.

Conversely, on other occasions, it is worth taking a little longer time to mask using a cheaper material. For instance, if using brown paper and tape to mask the side windows of a vehicle rather than pretaped sheet, it only takes a minute or two longer, but the brown paper is much cheaper. The vehicle in Figure 5.22 has had a lot of time invested in the making but using the cheaper materials option of brown paper.

Another concern is using the correct material for the job at hand. It has been witnessed on many occasions where refinishers have used huge amounts of the wrong masking material, rather than thinking the job through and using the correct material. The author has actually watched an experienced technician using multiple strips of 2 in. tape in order to cover an entire rear quarter window aperture, rather than outline with the tape, and then use brown paper to fill the void. Another frequently seen technique that wastes material is the use of tape to wrap a mirror or rear light in its entirety, rather than a small piece of brown paper or stat sheet. On occasion this cannot be avoided though, the headlight featured in Figure 5.23 may have caused all kinds of issues with masking moving around if it had been masked with paper.

Figure 5.22 There is a huge amount of time invested in the masking of this vehicle. The materials are used sparingly and neatly.

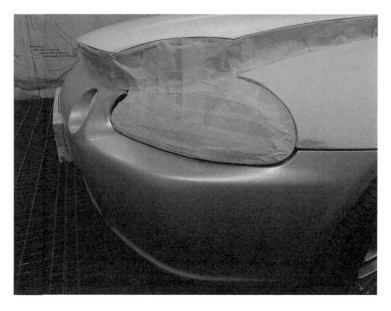

Figure 5.23 This headlamp has been masked using 50 mm tape as it needed to be painted with the bonnet open, so tape was being used to fill the void. It was necessary to mask using the tape in order to keep it tight and prevent it flapping around. This is not the most economical way but does provide a good clean area around the repair.

Below Figures 5.24 and 5.25 show an excellent example of economic masking. The vehicle has been completely covered with stat sheet which has been used right up to the rear, masking around the edge of the panels, and even into the void for the rear lamp. The wheel wells and sill cover have been masked with brown paper and tape – used sparingly, of course!

Figure 5.24 The masking sheet is used right up to the repair and shows an excellent use of materials.

Figure 5.25 The only paper to be used is covering the wheels.

In brief, when planning a masking job, consider which materials will reduce the labour time considerably, which ones will save money by basically covering a component quicker, and how much a masking material costs to buy in the first place. Finally, do not waste materials. One of the biggest costs that could be reduced when using masking materials is to use the entire roll of tape. It is often seen that technicians will use half-spent rolls of tape to hold open bonnets or as a stand for a small component to be balanced on. A good habit to get into is not leaving rolls of tape on the floor where they get run over. Masking tape should not be left on the floor anyway, as the tape can be easily contaminated with dust from the floor, which then gets transferred to wet paint film, resulting in dirt nibs.

5.15 MASKING FAULTS

Remember that masking products need to be accurately and carefully applied. The door handle in the image Figure 5.26 has been previously masked inaccurately. Where the tape touched the door skin, it has clearly left marks on the panel. As can be seen, the paint edge that has been created is an excellent dirt trap and will need careful sanding back and feathering in before the handle is carefully remasked or removed and the panel is repainted.

Figure 5.27 shows the same panel with another type of poor masking, where the component to be protected has not been properly covered. In this instance, the black metal window scraper has silver paint on it. This will be difficult to remove without damaging the silver paint on the door or the black finish on the window scraper.

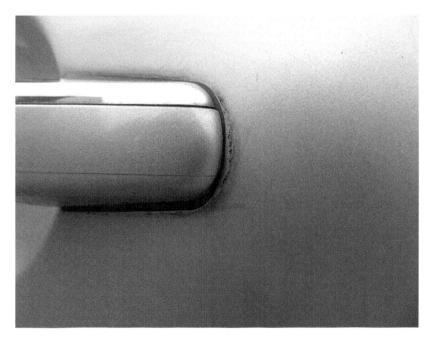

Figure 5.26 This handle has been previously masked badly, with the tape touching the door skin. This has left an edge which can lift and has dirt ingrained under it.

Figure 5.27 Here the tape was not fully tucked under the window trim, and as a result, silver paint can be seen on the edge of the black trim.

The panel shown in Figure 5.28 was left exposed, while the other side of the vehicle had white high-build primer applied. The resultant overspray shows as tiny white bumps on the surface. In order to remove these bumps, the panel will need to be flatted with P1500 followed by P2000, and then polished.

Figure 5.28 Due to poor masking, this panel has been contaminated with primer overspray.

Figure 5.29 Here top coat overspray from an adjacent panel repair has leaked through a gap in the masking and resulted in this hazy patch.

On the same vehicle, there is a different type of overspray. Here in Figure 5.29, there has been a leak in the masking which has allowed atomised paint to come through it and create an area of dry matt overspray. This can be easily removed with polishing alone.

Chapter 6

Paint application equipment

As with many processes, the tools required to apply refinishing paint materials are many and varied. It is therefore important to understand how these pieces of equipment are operated and how their internal components interact and fit together. In this chapter, the benefits and drawbacks of each piece of equipment will be discussed, as well as the more in-depth explanation of how the tools work. This should help technicians decide on which type of equipment will suit them best and make informed decisions when selecting new tools. Figure 6.1 shows a technician in typical pose spray gun in hand.

Figure 6.1 Applying paint is a difficult procedure where you are trying to apply a liquid to an often vertical surface without touching it.

6.1 APPLICATION TOOLS AND EQUIPMENT

The most obvious piece of equipment required to spray a vehicle would be a spray gun. Indeed, a refinishing technician's spray guns should be regarded as the tools with which they earns their living. Therefore, when choosing a new spray gun, there are several things to consider:

- Spray guns are designed to apply different coatings, and consequently, it is virtually impossible to find one gun that will apply all the different types of top coats required for the refinishing of motor vehicles.
- Don't buy the cheapest spray gun! There are many makes on the market, and generally speaking, it is worth spending the several hundreds of pounds required to obtain a good quality spray gun. It has been witnessed on several occasions that cheap spray guns initially produce an acceptable level of atomisation and fan pattern – however, after only a few uses, they produce less pleasing results and will often struggle to apply base coats smoothly and evenly.
- What types of vehicle are you likely to be working on? There is little point in investing the money for a pressure-pot system if you mainly perform SMART (small and medium area repair technique) repairs. Conversely, there is little point in buying a mini-style spray gun if you paint the sides of lorry trailers on a daily basis.
- Look after and maintain your spray guns. As has already been said, a good quality one costs several hundreds of pounds and, if looked after correctly, will provide years of service. The key point here is looking after your spray guns. They require thorough cleaning after each use. If paint is still visible on *any* part of the gun, then it is not yet clean. A full strip down (Figure 6.2) is not required after each use but should be regularly performed, the regularity of the strip downs depends on how often you use the gun. Before undertaking this task, make sure you know how all the components go back together.

6.2 TYPES OF SPRAY GUN

6.2.1 Mini spray gun/touch-up gun

Small spray guns have always been available in suction- and gravity-fed, but they are now predominantly gravity-fed (Figure 6.3). They are particularly useful when painting complicated components, such as the spokes of aluminium wheels and mirror covers, as they can be manoeuvred more easily and fit into tighter gaps. They are often used by S.M.A.R.T. technicians for partial panel repairs and blow-ins, as their smaller size obviously gives a smaller fan pattern, thus enabling the repairs to be kept local.

6.2.2 Gravity-fed guns

These have been by far the most popular spray guns in use in body shops for many years, and this is due to them offering simplicity of design and being lightweight in use. The simplicity has several benefits, including ease of cleaning, as the passage the paint passes through is very short, which obviously means there is less to clean up.

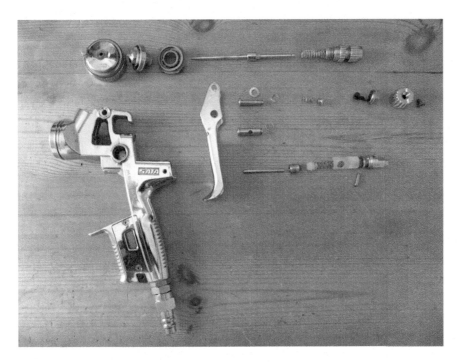

Figure 6.2 Spray guns can be broken down into their component parts very easily. However, they are easy to damage, so it is important to know what you are doing.

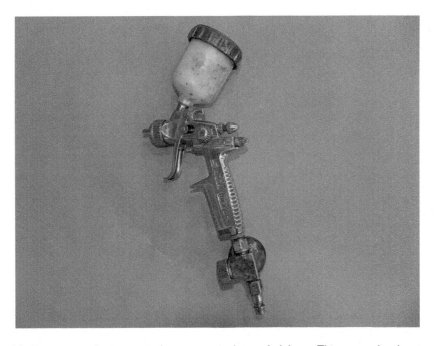

Figure 6.3 Mini jets are the same as other spray guns but scaled down. This means they have smaller fan patterns that make them well suited to small repairs.

There are many excellent gravity-fed guns to choose from (Figures 6.4, 6.5 and 6.6), and some have further advancements, such as built-in digital pressure gauges, self-heating bodies, or even disposable air caps and fluid tips.

The main drawback of using gravity-fed spray guns is that the majority of the weight is above the technician's wrist. This means the gun will want to tip if it is balanced on the trigger guard, unlike a suction-fed spray gun that has most of the weight low down. However, most technicians find that the gravity-fed spray guns are easier to manoeuvre, despite this drawback.

6.2.3 Suction- or syphon-fed guns

At one time these spray guns were seen in the majority of body shops, to the point that many people believe they are an older design. They are far less regularly seen now, with gravity-fed guns being the more popular for the past 25 years or more. Gravity-fed (Figure 6.6) and suction-fed guns (Figure 6.7) are, however, contemporaries when considering which is the older design. They offer a more balanced feel when being held, as all of the paint (and weight) is below the operator's hand, so when the gun is held with one finger behind the trigger, it will balance easily. They also offer a larger paint pot, which is why they are now more often seen in large vehicle refinishing operations, such as HGVs Heavy Goods Vehicles. The main disadvantage is that they are more difficult to clean, although currently there are a few manufacturers making disposable mixing cup/paint pot systems for them.

Figure 6.4 There are many manufacturers of gravity-fed spray guns. They all work on the same principle, but the controls will be laid out slightly differently.

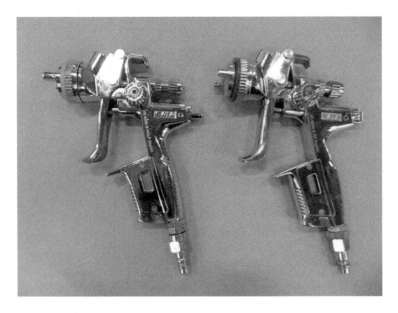

Figure 6.5 These two spray guns at first appear identical; however, they have very different setups that apply the paint differently – one for base coat and the other for clear coat.

Figure 6.6 Each manufacturer will design their guns to apply paint in different ways. As a result, refinishers will have their favourite gun manufacturer simply as it suits their style best.

Figure 6.7 Suction-fed spray guns have the paint reservoir below the gun body. This gives excellent balance and the opportunity for a larger capacity for paint.

The reason why it is more difficult to clean up these guns is due to their suction (or syphon) tube, which is a metal pipe approximately 200 mm long, which sucks the paint out of the paint pot, and as a result, it requires thorough cleaning both inside and out. When the trigger of a suction-fed gun is depressed, air passes through the internal passages and over the top of the syphon tube. This creates a negative pressure, which causes the paint to be sucked up the syphon tube, where the two constituents (air and paint) are mixed, and then delivered to the atmosphere via the fluid tip and air cap.

As the air and paint are mixed together after being emitted, suction-fed spray guns are known as 'external mix' spray guns. There are variants of suction-fed spray guns where the air and paint are mixed internally, but these are very much for the DIY market and will not be discussed here.

6.2.4 Pressure-fed guns

Pressure-fed spray guns (Figure 6.8) are used mainly for big jobs such as full repaints on large cars and vans, or repairs to HGVs. They are basically a modified suction-fed gun, with the pot being mounted remotely and with the syphon tube being extended, lengthened and swapped to a rubber or flexible plastic hose. The advantage of this is that the technician does not need to hold the pot at arm's length, so it can be made much larger.

The pot can be hooked on the air-fed belt if it is still reasonably small (generally around 1.5 litres), or left static on the floor of the booth, and moved around if larger. Obviously, when mixing two-pack paint materials, the pot life should be taken into consideration before filling a very large pressure-fed pot, and the painter should be confident that they will finish the job before the material 'goes off'.

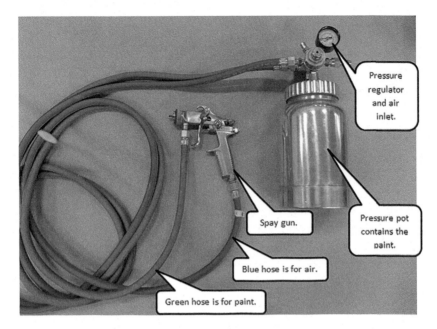

Pressure regulator and air inlet.

Spay gun.

Pressure pot contains the paint.

Blue hose is for air.

Green hose is for paint.

Figure 6.8 **The pressure pot system is a modified suction-fed spray gun. Air enters the top of the reservoir and pressurises the pot. It then continues up the blue pipe to the spray gun. The pressurised paint travels up the green hose to the gun.**

Besides the spray gun having a normal compressed air hose which is used to atomise the paint, the pressure-fed pot is also supplied independently with compressed air, which then pressurises the system. This pressure forces the paint up the delivery hose to the spray gun.

Like the suction-fed gun, cleanup is much more complicated than a gravity-fed gun, and a pressure-fed gun also has the added complication of the long fluid delivery hose which needs to be kept spotlessly clean. Due to this problem, some workshops use disposable delivery hoses to save on lengthy cleanup which would use large quantities of gun wash.

6.3 PRESSURE GAUGES

In order to apply paint in a consistent manner, a pressure gauge should be used. There are many different types of gauges. At least two spray gun manufacturers are making guns with inbuilt digital pressure gauges. Figure 6.9 shows a gravity fed gun with a small digital gauge in the base of the grip. These help keep spray guns lighter and more manoeuvrable. However, they do need to have their batteries replaced regularly.

Many manufacturers produce separate digital pressure gauges which may either screw to the bottom of the gun or alternatively locate on the rear of the gun body above the grip and below the paint pot. The most typical type is an analogue gauge, screwed inline to the air inlet at the bottom of the gun, a well used example of which can be seen in Figure 6.10. These should be screwed on tightly to prevent air leaking which would give a false pressure reading.

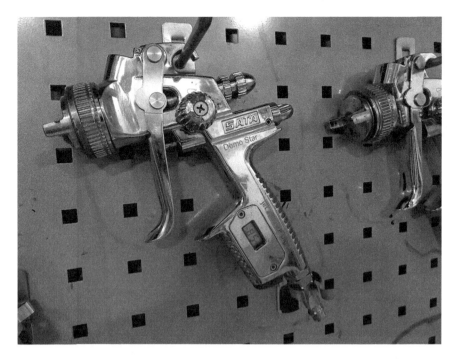

Figure 6.9 This gun has an inbuilt digital pressure gauge. The battery needs changing, hence the false pressure reading visible in this picture.

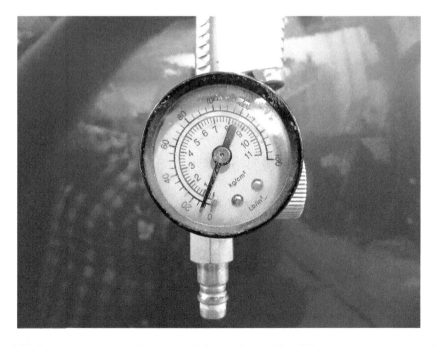

Figure 6.10 Analogue gauges can be screwed inline at the air inlet. This one measures in psi in black and bar in red.

It should be remembered that the air pressure indicated on the gauge is at the gun inlet, not at the air cap. Pressure at the air cap would be more accurate but would necessitate extra equipment which could not be sprayed through. It is therefore commonly accepted that pressure at the gun body is an adequate measure, and figures given in paint manufacturer's technical data take into account the slight pressure drop in the spray gun.

Analogue pressure gauges have an advantage of reading in both bar and psi (pounds per square inch), whereas digital gauges typically only read in either bar or psi. It is usual for refinishers in Europe to use bar; however, technical data for paint often gives the desired pressure in both units.

In theory, every type of paint could have its own preferred application pressure, although realistically, most materials have a recommended application pressure of 2 bar. Having said that, there are many instances when the pressure may be different, such as performing 'drop coats' on metallic base (typically 1.5 bar) or applying polyester primer (typically 3–4 bar).

When setting the pressure of a spray gun, the trigger should be held at the halfway point, where air is allowed to exit the gun with no paint, and the airflow knob should then be adjusted to the correct pressure, while the air is passing through the gun.

6.4 DISPOSABLE PAINT CUPS

One of the best accessories for spray guns in recent years has been disposable paint cups (Figure 6.11). They offer the advantage of built-in filtration, so there is no need for straining paint after mixing. They also remove the need to clean up the reusable paint pot supplied with spray guns.

As the mixing cup becomes the paint pot, there isn't any wastage incurred when pouring paint from one pot to another, and every job also starts with a new clean material cup on the gun, as can be seen in Figure 6.12.

Figure 6.11 Disposable material cups combine the mixing cup with a filter and a lid which can then be used on top of the spray gun.

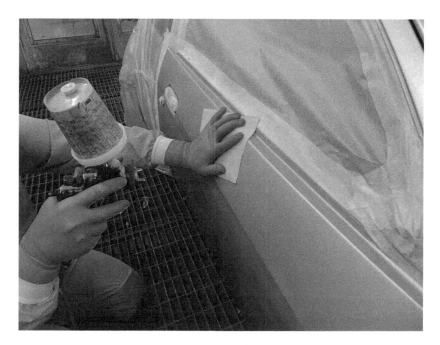

Figure 6.12 Some manufacturers make collapsible material cups that compress as the paint is used. Others use hard cups that need an airway entering them in order to allow the paint out of the bottom.

Some designs have a hard outer cup which may include mixing ratios and volumes in millilitres printed on it or come with an attachment to help measure either ratios or volume. This type of disposable cup holder will have a soft liner which will crush as the paint is vacuumed from the cup while spraying. This has the advantage that the gun will still spray while upside down, as the paint is always held at the orifice of the paint passage. This feature is especially useful under tailgates, wheel arches, sills, etc. Despite these advantages, some technicians are still put off by the fact that the outer cup has to be reused.

As a result of this, some manufacturers offer an alternative, which uses the hard cup as the disposable component. This design also has ratios and millilitres marked on the cups. There is a disadvantage to this simplified design though, which is that there needs to be a way for air to get in to replace the paint coming out. This means that the pot has a hole in the top, and it will leak if the gun is inverted, so it is not possible to spray for lengthy periods while holding the gun upside down.

Disposable paint cups are available for gravity-fed and suction-fed guns. Most of the manufacturers of these cups offer them with different capacities, so there is something suitable for different job sizes. As they are made of lighter material than the conventional pots, they are also slightly more comfortable to use. All of the manufacturers use variations of a bayonet or a coarse thread fitting, and each will have a different adapter to fit the spray gun – apart from a few that are designed for specific spray guns.

Disposable paint pots can be fragile and easily torn, so they should be used in conjunction with plastic disposable mixing sticks. These have the added bonus of being clean for each mix, as they are one use only. They are also shaped with a relaxed 'S' or 'Z' profile, which can be clearly seen in Figure 6.13 this makes scraping the side of the pot much more thorough (Figure 6.14). This ensures that all of the pigments are incorporated into the mix, which is vitally important for accurate colour matching.

Figure 6.13 Disposable plastic mixing sticks have an S or Z profile to help scrape tinters off the side of mixing cups.

Figure 6.14 It is necessary to use plastic mixing sticks to prevent damaging delicate mixing cups.

6.5 ANATOMY OF A SPRAY-GUN

Spray guns – no matter whether they are large or small, gravity-fed or suction-fed – have basically the same components. Figure 6.15 shows an example of a gravity-fed spray gun with all of its components (except the digital pressure gauge) removed.

It is important to understand how these components work together in order to keep your spray gun operating correctly. This is especially important when your gun is not working properly. It is therefore essential to know how to strip a gun without damaging it and how to rebuild it in the correct order. It is also very useful to be able to name the components correctly, so that the correct replacement parts can be ordered.

It is not normally necessary to strip a spray gun to the extent of the one in Figure 6.15, unless there is a suspicion of contamination or a broken component. In order to strip a spray gun to the extent of this, you will need only simple tools, such as spanners, allen keys and screw drivers. Most manufacturers of spray guns supply their products with a combination spanner and any specialist tools required like those seen in Figures 6.16 and 6.17.

Number	Component name
1	Air cap
2	Gun body
3	Fluid control knob
4	Fluid needle spring
5	Fluid needle
6	Fluid tip
7	Baffle
8	Trigger
9	Air valve push rod
10	Air valve retaining screw
11	Air micrometer
12	Trigger sleeve.
13	Trigger pivot and retaining clip
14	Gland nut, packing and spring
15	Spindle for fan control
16	Fan adjustment knob

6.6 STRIPPING AND REFITTING SPRAY GUN COMPONENTS

The spray gun in the Figure 6.18 has been stripped down to the 'light service' stage. Many technicians will do this daily, in order to ensure that no paint from previous jobs remains in the fluid passages.

The first component to strip is the air cap ①. This is a threaded component and unscrews like a bottle cap.

At the opposite end of the gun body ② is the fluid control knob ③, which also unscrews by hand. Be careful not to let it fly off, as it is being pushed out by the needle return spring ④.

Once these two items are safely stored, pull out the fluid needle ⑤. Some technicians remove the fluid tip ⑥ prior to the fluid needle; however, this can 'grind' the two

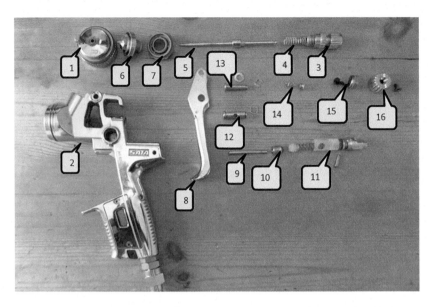

Figure 6.15 Ensure you keep a track of where all the components come from when stripping spray guns.

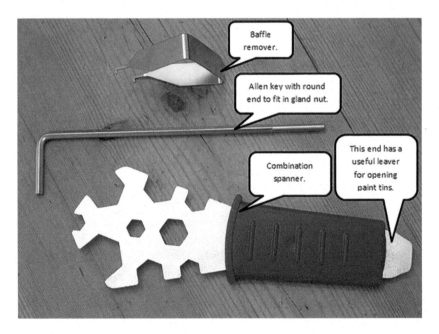

Figure 6.16 Spray gun manufacturers supply the tools specific to their own guns.

components together, causing premature component wear, so it is recommended to remove the fluid tip after the needle.

Once all of these components are removed, they should be thoroughly inspected, then cleaned with paper towel and the correct solvent for the paint used.

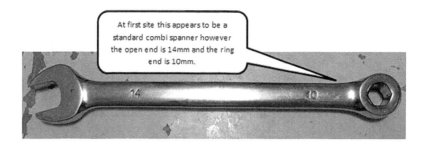

Figure 6.17 This unusual spanner is all that is required for the spray gun it is supplied with.

Figure 6.18 This spray gun has been partially striped. Many refinishers will strip a gun this far after each job, although this isn't always necessary.

When reassembling from the light service stage, it is just as important to observe the correct assembly sequence. The first component to go back on is the fluid tip, and then the needle. It is done in this order for the above reason of not grinding the two components together.

After this, it is simply a case of reassembling the spring and fluid control knob, and then the air cap. It is best practice to lubricate all threads prior to reassembly with a dedicated gun grease, which must not be allowed to leach into the paint passage.

6.7 SPRAY GUN SETUP

The three components featured in Figure 6.19 are referred to as the gun 'setup'. They should always be fitted as a matched set and are supplied as such by the spray gun manufacturers.

Figure 6.19 **These three components make up the setup of a spray gun.**

The air cap ① in the image is labelled as 1.2, slightly above the main aperture. This signifies that it is designed to be used with a fluid tip ⑥ which has an orifice of 1.2 mm diameter, and this in turn must be used with a fluid needle ⑤ intended to fit into the 1.2 mm hole. Once the needle and fluid tip have been used together, they should not be exchanged with other components as they will wear together, meaning a drip-free seal is maintained.

6.8 HIGH-VOLUME LOW-PRESSURE VERSUS COMPLIANT SPRAY GUNS

Before the advent of high-volume low-pressure (HVLP) spray guns, there were conventional spray guns, which were designed to atomise paint into an extremely fine mist, then propel that mist onto the panel at great velocity. This way of applying paint gave excellent finishes but was particularly uneconomic in paint consumption.

The idea behind HVLP is to use a larger quantity of air at a lower pressure at the air cap of the gun. This has the effect of atomising the paint into slightly larger droplets. These droplets are then pushed towards the panel less aggressively. Due to this effect, there is much less wasted paint caused by the paint either missing the panel or, more importantly, being lost through the droplets bouncing off the panel due to the high velocity and eddies caused by the air movement.

In order to help one visualise the difference between the two types of spray gun, consider a conventional spray gun to be like a hose pipe with a small quantity of water running through it, and the end is restricted by one's thumb. This would produce a thin stream of water that would hit a small area of a panel and then spray all over the place.

Now consider the HVLP spray gun to be like a nonrestricted hose pipe passing a larger quantity of water onto a panel. This would soak a large area of the panel without the same type of bounce back.

To confuse matters, the spray guns that can be seen in Figure 6.20 are both HVLP, but designed to apply different finishes. The comparison between the appearance of the clear coat gun on the right and a conventional spray gun can be made. However, it must be remembered that both of the spray guns pictured are HVLP.

In order for a spray gun to be considered an HVLP, it must have a transfer efficiency of at least 65% compared to only 20%–40% for a conventional spray gun. 'Transfer efficiency' is the measurement of the weight of the solid material within

Figure 6.20 Compare the gap between the air cap and the fluid tip for the base coat gun on the left and the clear coat gun on the right.

the paint that has ultimately adhered to the panel. Many spray gun manufacturers have now designed spray guns that far exceed the above minimum requirement. Some manufacturers have produced spray guns that are not true HVLP, but they are more than 65% efficient.

6.9 BOOTHS & DRYING EQUIPMENT

There are several ways in which paints can be force dried. These include the use of hot air, such as that in spray booths and ovens, infrared and near-infrared heat lamps, increased air flow and UV radiation.

Depending on the type of paint being applied, it will need to be dried in different ways:

- Traditional single-pack top coats dry by solvent evaporation and, as such, need warm, dry conditions for their optimum drying environment. Therefore, booths designed for the application of single-pack paints do not require a 'bake' function, but simply warm air and adequate extraction to remove the off-spray.
- Two-pack materials cure through polymerisation, which is a chemical reaction of molecules cross-linking. This can be sped up by heating the paint to around 60°C–70°C, which can be achieved by several means.

- Water-based paints, being single-pack, require only increased warm air movement to promote drying through evaporation of the water molecules.
- An underused technology is the type of paint that cures very quickly due to exposure to intense ultraviolet (UV) light. The UV light can be provided by either sunlight or, more reliably, artificial light.

6.10 SPRAY BOOTHS

Historically, spray booths were designed to simply provide fume extraction and warm dry air for an ideal environment in which to apply and dry paint. Nowadays, most spray booths also have a bake cycle included to speed up the drying of two-pack materials. Indeed, there are many technicians who insist that two-pack paints must be baked in order to cure completely.

Spray booths (an example of which can be seen in Figure 6.21) can be fuelled by oil but are much more often heated with indirect gas burners. Most have the ability to be adjusted to provide different lengths of bake time, as well as accurate temperature increments. Many booths have touch screens that allow the bake cycle to be adjusted to slowly increase the temperature over many minutes, if that is what is desired.

It is usual to bake plastic components such as bumpers at cooler temperatures of around 60°C when they are off the vehicle placed on stands. This lower temperature will prevent them from losing their shape in very warm bakes.

Figure 6.21 The spray booth provides a clean environment in which to paint the vehicle, as well as extracting the foul air through floor filters in order to make it safer. The extracted air is then expelled to the outside environment.

With the advent of volatile organic compounds (VOC) compliance in 2007 came the general acceptance of waterborne base coats, which offer many excellent features. However, due to the use of water as the solvent, the drying times compared to solvent base coats are predominantly much greater. For this reason, booth manufacturers were prompted to include drying facilities directly aimed at dramatically increasing airflow, while keeping temperatures moderate.

Waterborne base coats require this increase in airflow for efficient drying times, and although warming the booth up by 10°C–20°C does help dry the paint faster, baking at 60°C–70°C can lead to further problems, such as high panel temperature preventing the clear coat that follows from flowing out and creating a smooth finish. These high temperatures do not reduce drying times dramatically either.

6.11 HANDHELD AIR DRIERS

For smaller areas of repair, technicians may opt to use a small air drier as shown in Figure 6.22. These small driers can be easily held by the refinisher and directed to the required area. Alternatively, they can be placed on a stand and left to do their job while the clear coat is being mixed. The advantage of using a handheld drier such as the one in the image is that it can be used to help base coat flash-off between coats. They can also be used to dry the final coat if the technician is not pushed for time. If the technician is in a rush they may wish to place the air blower on a stand.

The advantage of using a stand for air driers is, several air driers can be directed at a panel at one time. Not only does this mean a larger area can be dried at once, but the actual air movement is increased, as the air jets can be positioned at opposing angles, thus creating eddies and vortexes.

When using these small driers, it is very important to understand how they promote evaporation. The idea is to angle the jet of air across the wet paint to encourage the water molecules to evaporate out of the paint film. The advantage of using a dedicated air drier opposed to the air from the spray gun is that the air leaving the drier is warm. The compressed air leaves the handle and exits the barrel forward. This then pulls warm air from the booth through the blow gun, which is then directed to exactly where it is required. This is a very economical way of drying single-pack paints.

It is imperative not to use an air drier on two-pack materials. This is because the air movement causes the paint to 'skin' over; this then slows down the polymerisation, which is the chemical reaction that causes these materials to cure.

It is advantageous for two-pack materials to cure from the inside out. As well as polymerisation, they also require the solvents to evaporate from the film, so in the situation of the paint skinning over, the solvent will burst through the outer skin, leaving little craters, technically referred to as 'solvent pop'.

6.12 ANTISTATIC BLOW GUNS

Although not actually drying equipment, antistatic blow guns (Figure 6.23) are very similar to handheld driers, except they are used prior to applying paint rather than after. They are used to remove static from panels (Figure 6.24) in order to give a less patchy finish with less dirt nibs attracted.

Figure 6.22 Handheld air dryers speed up the drying times of water-based paints by encouraging the water to evaporate. They must not be used on two-pack materials as they will promote solvent pop.

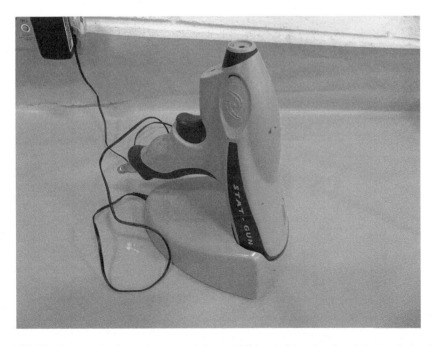

Figure 6.23 The Stat-gun is charged on a special stand. This provides the electricity needed to ionise the atoms emitted.

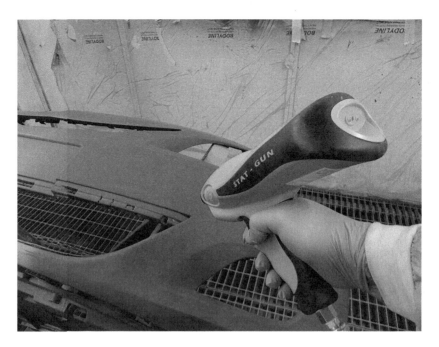

Figure 6.24 The antistatic blow gun is especially useful on plastic components around panel transitions that may not have been rubbed with the tack rag.

The antistatic blow gun passes electricity through the air passing from it, which then balances out any static on the panel caused by cleaning with panel wipe rags or tack rags.

6.13 INFRARED HEAT LAMPS

A useful alternative to heating by gas or oil is to use electric infrared heat lamps. These can be portable such as the ones in Figures 6.25 and 6.26 or they can be fixed into a moveable arch, which then moves through the booth much like an automated car wash does.

The advantage of using portable heat lamps is that they can be used anywhere that has mains electricity, so they can free up booth time when drying smaller repair areas. They are most often used for drying patch-primed areas in the workshop. These could be roll-primed or sprayed in the booth and then moved into the workshop for drying.

Heat lamps are not restricted for use on primer. They can be used for baking top coats, although it is not recommended to move a vehicle that has wet top coat on it, so they are unlikely to free up booth time particularly well. Obviously, if the top coat is on a smaller panel, such as a wing or a wheel, it makes much better economic sense to move the component out into the workshop to bake under a heat lamp, rather than heat up an entire booth in order to dry such a small component.

Although it is true that infrared lamps will speed up the drying of waterborne base coats, it is quicker to dry them with the aforementioned air blowers.

Infrared lamps warm panels through radiant heat. This means that they have the advantage of heating up the panels underneath the paint, making infrared particularly useful for two-pack materials. This is due to the warm substrate helping to evaporate the solvent and speeding up the cross-linking that takes place in two-pack materials in order to achieve a full cure. As well as being suitable for top coats, infrared heating can be used to very good affect on filler materials.

6.14 UV (ULTRA VIOLET) CURING

In recent years, paint manufacturers have tried several technologies to improve the curing of paints. There have been some very unusual systems that have not taken off. However, UV curing has a degree of popularity that is predicted to increase with time.

There are two clear advantages of using UV light to cure paint:

- The first of these is the speed at which UV products dry; full cure can be achieved in just a few minutes.
- The second advantage is the economical use of energy. UV lamps can be directed precisely at the panels that have been refinished, negating the need to heat an entire booth.

Figure 6.25 Smaller heat lamps like these are very useful for curing the paint on the sides of vehicles.

Figure 6.26 This type of heat lamp which is on an adjustable arm is especially useful for horizontal panels such as bonnets and boot lids.

On the other hand, there are several disadvantages that appear to be preventing the technology being taken up by more body shops:

The first of these disadvantages would be the initial setup cost of purchasing the lights. This could be overcome by positioning the panel or vehicle outside and using natural sunlight. Obviously, this isn't so easy if the weather is cloudy or wet, and as has been previously mentioned, vehicles with wet top coats on them should not be moved.

There have also been a few health scares surrounding UV light. The manufacturers have worked hard to make the light intensity required for curing the paint as safe as possible. Of course, as with all paint materials, it is recommended to wear the personal protective equipment (PPE) advocated by the paint manufacturer.

Another drawback is the possibility of accidentally curing the material while it is still in the gun.

In conclusion, the selection of paint application equipment is a very personal choice. There will be many technicians who claim that *their* favourite make of spray gun is the best, and perhaps for them it is. However, it is recommended to trial as many as possible before making the costly outlay for one.

This chapter is meant to give a nonbiased opinion in order to help refinishers make informed decisions on the purchase of their spray guns. Furthermore, the explanation of drying equipment is intended to help prevent confusion with which type of paint drier to use in order to gain the best possible final finish.

Chapter 7

Application of top coats

The application of top coats is the area where most skill is involved – after all, a refinishing technician is trying to perform the seemingly impossible task of making a liquid stick to an often vertical surface without touching it! What makes it even more difficult is that the technician needs to take into account panel contours, colour and textures. This may mean blending colour into surrounding panels or attempting to achieve a glass-like finish with the clear coat. Bearing this in mind, the only way to become good at putting paint onto a panel properly is to practice. This is of the utmost importance, because if top coats are applied too heavily, not only can they run, but they may also change their value (shade). If they are applied too lightly, the finish will have an unpleasant 'orange peel' texture. In order to help make this job easier, there are several precautions, techniques, tools and equipment that can help.

7.1 HOW TO CHOOSE REFINISHING TOP COATS

A refinishing top coat is any paint material that is seen after the vehicle paint application is finished. This basically means base and clear coats, and direct gloss coats. There are some specialist coatings that also count, such as translucent pearl coats, tinted clears (often referred to as 'candies') and metal flakes.

It is usual to repair like with like, so original equipment manufacturer (OEM) base and clear should be repaired with refinishing base and clear (as in Figure 7.1), and factory direct gloss should in turn be repaired with refinishing direct gloss. Having said this, it is also common practice to repair direct gloss finishes with base and clear coats, as the base coat is easier to blend, and the clear coat is easier to blow in.

7.2 DIRECT GLOSS

Direct gloss paints may be single-pack or two-pack. They are paints that provide colour combined with opacity, and dry or cure with a full gloss. They are often referred to as 'solid' finishes, as there is no clear coat over the colour.

It is now unusual to find single-pack direct gloss paints in use for refinishing or by OEM vehicle manufacturers. This is due to the resultant dry film of single-pack paints being far less chemically stable. The poor stability of the dry films means they often lose gloss faster, and they have less colour-fast pigments compared to two-pack

Figure 7.1 Top coat is applied as per the manufacturers' recommendations. In this case, full coat followed by three-quarter coat and then a drop coat.

materials. They are also more susceptible to attack from spilt solvents, such as petrol around the filler neck.

Single-pack paints are usually higher in volatile organic compounds (VOCs) than two-pack materials, because so much of the wet film is made up of thinners. The paint, before mixing, contains thinners, and then it is mixed for application with up to an additional 50% thinners.

Thinners is the component contributing most of the VOCs found in paint materials. As single-pack materials dry through evaporation of the thinners, it makes them worse for the environment, as ALL of the thinner vapors are ultimately released to the atmosphere. These reasons all add up to single-pack direct gloss being far less popular, so it is therefore recommended to use two-pack materials as far as possible.

Direct gloss paints are often used on commercial vehicles, as well as some passenger cars. They can provide the same level of protection and pleasing appearance as base and clear, but as there is one less application process, it can work out more economical in both material costs and labour times. During repairs, the same savings can be made, so commercial vehicles are often repaired with direct gloss.

When choosing to use direct gloss, the repairs are most often undertaken as edge-to-edge repairs, which means panels are painted from one edge to the other. In order to achieve an undetectable repair, it is essential to match the colour perfectly. This is not the only hurdle, as even two-pack direct gloss paints fade and lose their gloss quicker than base and clear.

So, even if the colour match is mixed using the exactly same pigments as the OEM, it may well still result in a mismatch. This will be due to fading of the original finish, and then this is further magnified with loss of gloss, as reflections play a huge part in colour matching. If the original panels are suffering with loss of gloss, this can be rectified by polishing the panels adjacent to the panels to be repaired.

Having described the drawbacks of edge-to edge repairs, it should be pointed out that direct gloss paints *can* be faded out, through either the repaired panel or adjacent panels, and this will be covered later in this chapter when describing top coat application techniques. Direct gloss is more often than not nonmetallic (as pictured in Figure 7.2); however, metallic direct gloss is available. This can be very difficult to apply and is only really used when repairing vehicles from the1970s or 1980s, as occasionally these older vehicle colours have not had the original direct gloss formulas converted to the base-coat schemes.

In summary, direct gloss paints offer greater economy in both material and labour costs. It is also slightly easier to prepare the panels for the top coat, as it has a greater film thickness than base coat, and consequently slightly coarser grades of abrasive can be used than when preparing for clear-over-base (COB) systems. However, the COB system offers ease of blending, along with longevity of both pigment and gloss.

Figure 7.2 Direct gloss offers an excellent clarity of colour, which particularly suits older vehicles and bright clean colours.

7.3 BASE COAT AND CLEAR

Also known as clear over base (COB) (Figure 7.3), base coat can be solvent-based base coat or water-borne base coat, while the clear coat can be polyurethane, acrylic, or UV cured. The current industry standard for base coat is waterborne, although solvent-based is still available for specialist coatings. This may be for custom paint work, or for use in vehicle restoration, where some owners insist on their vehicles being repaired with the original type of paint.

It is recommended to choose waterborne base over solvent-based for standard crash repair procedures, because it has far lower VOCs. Waterborne base coats easily offer the same colour diversity as solvent-based. There are occasions though, when solvent base can offer greater robustness, such as for wheel refurbishment, but waterborne base is being continually developed to offer equal properties to solvent base.

Early waterborne base coats would take much longer to dry than their solvent-based counterparts, but with constant development, water-based now dry much faster. The hydrocarbon-based solvent in solvent-based base coat is much more volatile, and there-fore, it still dries much faster than water-based.

When undertaking custom paint work with many tape-outs or stenciling work such as those seen in Figure 7.4, solvent base coat is the best choice. This is because it can offer a greater ability to withstand the application and peeling of masking tape and stencil film.

Most paint material manufacturers supply acrylic two-pack clear coat, which offers excellent durability against UV fading, chemical attack and abrasion from general use. The current industry standards for clear coat are high solids (HS) and

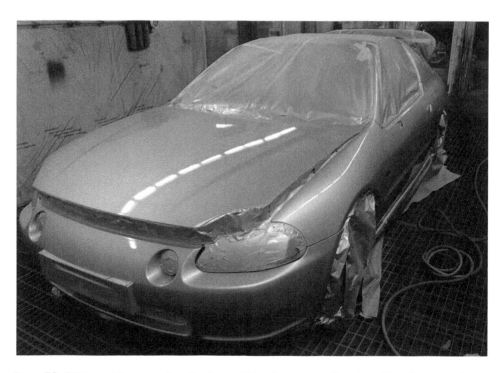

Figure 7.3 COB provides ease of application and blending on metallic colour. The gloss immediately after application is apparent here.

Figure 7.4 When undertaking complicated tape-outs like the ones on this bonnet, solvent-based base coat offers greater durability against glues, sanding and panel wipes.

ultra-high solids (UHS). These types of clear coat offer greater film build than the more traditional medium solids (MS). They also have much lower VOCs, which is a necessary improvement in order to comply with current legislation regarding the control of VOCs.

Single-pack clear coats are also available; however, they are not used in most repair shops due to the resultant high levels of VOCs. As with single-pack direct gloss paints, they can suffer with poor gloss and invariably have poor chemical resistance.

7.4 APPLICATION OF TOP COATS

Generally, it is recommended that the corresponding technical data for the paint being used be consulted for the manufacturer's recommended forms of application, as every manufacturer will offer very different instructions. When applying top coats, there is really nothing like experience to help a technician gain a good quality finish. However, there are a few simple rules to remember that will help *all* refinishers apply consistently good coats of paint, and these tips can be applied to all types of top coats. They are not listed in order of importance – they are all equally as important as each other.

7.5 PAINT AND PANEL PREPARATION

As important as all the other following points are, even the most experienced refinishers will struggle to gain a good finish if the paint has been mixed poorly, or to the wrong ratio. Therefore, all materials should be stirred thoroughly before and after mixing, and ensure that the paint is mixed with its specific thinner or hardener, at the ratio specified in the technical data for that product.

Use only new, clean mixing vessels. Disposable material cups (an example of one type can be seen in Figure 7.5) are very useful, as they provide a clean mixing vessel and a clean material cup each time; besides, they often also have built-in filtration. Obviously, if one is not using disposable material cups, the paint should be strained into the spray gun cup after mixing. It would be pointless ensuring that the paint is contaminant-free, and then applying it over panels with dust, oils and other contaminants on them. The vehicle should therefore be thoroughly panel wiped several times before the first application of top coat. It has been observed that, to undertake this to an acceptable level, the vehicle should be panel wiped before and after masking then again prior to the final tack-off.

Panel wiping should be undertaken with a panel wipe towel soaked with pre-cleaner (Figure 7.6) and a dry cloth to follow, and wipe off the panel wipe before it evaporates (Figure 7.7). When tacking off, it is best practice to blow the panel with the air from the spray gun (or an antistatic blow gun), slightly ahead of the tack rag. This process can be seen in Figure 7.8. This, of course, should be done while wearing gloves, ensuring they are clean and don't have wet paint on them from the paint mixing, which could end up being smeared across the panel.

At this point, there is an excellent opportunity to check all of the masking around the edges of the panels.

Figure 7.5 The disposable mixing cups have many ratios on the sides. There is also the useful feature of the filter built in to the lid.

Figure 7.6 Wet one paper towel with the panel wipe.

Figure 7.7 Use one wet paper towel and one dry paper towel. The panel wipe should be immediately
wiped off the panel with the dry towel before it can evaporate.

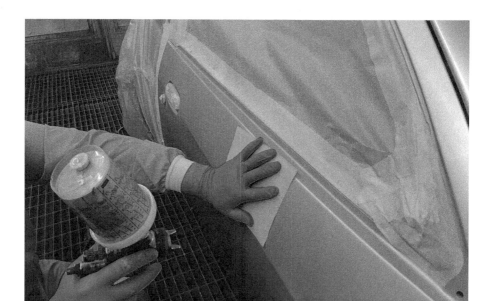

Figure 7.8 Following the panel wipe, the panels should be tack ragged with the tack rag in one hand and the spray gun blowing air in the other hand. The masking surrounding the repair should also be tacked in order to remove any dust or loose fibres which may fall into the fresh paint.

7.6 SPRAY GUN SETTINGS

Now that the panels and paint are ready, check the spray gun is set up not only to the technical data recommendations but to the refinisher's preferences – within the parameters given in the technical data. The air pressure can be set by pulling the trigger about halfway and should be checked after the fan pattern is adjusted to the required size; more often than not, this is full size or very near to it.

The fluid control should then be set to the refinisher's preference. It is usually recommended to go to as near to full flow as possible; however, the atomisation can often be negatively affected when set up like this, which is why the next step is vital: to perform a test spray, but *definitely not on the booth wall!* A proper spray-out sheet can be used, but where this is not available, a sheet of masking paper will suffice. This could be on a fully masked area of the car. The test sprays in Figures 7.9 and 7.10 have been performed on dedicated spray-out sheets.

When performing this test spray, the final settings should be altered to give the best possible fan pattern for the job at hand. It has been witnessed on several occasions where a refinisher has performed this test spray and found fault with the atomised fan pattern, then continued to paint the panel without rectifying these issues. That, of course, is not the correct procedure.

The test spray out can tell us if there is a problem with the paint mix and viscosity, or if there is a spray gun setup fault.

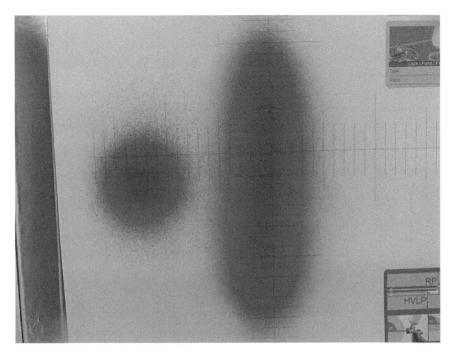

Figure 7.9 Direct gloss setups provide finer atomisation in order to achieve a smoother final finish.

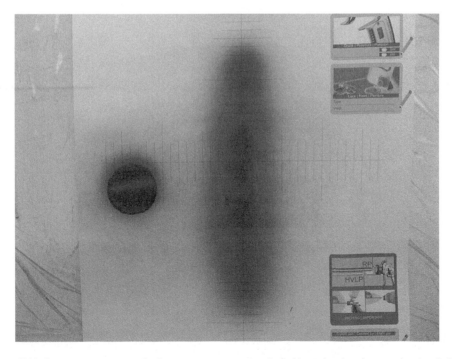

Figure 7.10 Base coat setups apply the paint in a more gentle fashion; the droplets tend to be slightly larger in order to provide an even coat with consistent opacity.

7.7 GUN DISTANCE

Simply put, the correct distance between the air cap and the panel for applying most paint materials is 15–20 cm (6–8 in.). Instead of carrying a ruler into the booth, a good starting point is to extend one's little finger and thumb to full reach, place the little fingertip on the panel and, with the spray gun held in the other hand, place the centre of the air cap on the tip of the extended thumb as in Figure 7.11. While this serves as a useful guide, the *real* skill is to maintain that distance constantly across the entire panel, taking into account the various contours.

When performing 'drop coats' on metallic base coats, the distance between the gun and the panel should increase to around 30 cm (12 in.), in order to allow the droplets to partially dry prior to landing on the panel.

There is a technique which can provide excellent results with clear coat. Where the gun is held much closer at around 10 cm, in order to gain a good finish, the gun speed must be increased dramatically due to the closeness to the panel. Remember that as the fan pattern is reduced, the number of passes has also to be increased.

7.8 GUN SPEED

It is very difficult to describe in words a speed at which refinishers should move the spray gun, and even more difficult to try and replicate it, as it is practically immeasurable. However, it has a massive impact on the resultant finish, and can be easily and immediately adjusted if the technician is not happy with the finish as it is being applied.

Figure 7.11 Checking spray gun distance.

The gun speed is directly linked to both the distance from the surface and the overlaps: if one of the three increases, one of the other two has to decrease. For example, if the gun speed is increased, the distance from the panel should be reduced.

7.9 ANGLE OF GUN TO PANEL

During application, the spray gun should normally be held perpendicular (at 90°) to the panel, both horizontally and vertically. If a spray gun is not held perpendicularly in the horizontal plane, it is likely to cause dry spray on gloss products and patchiness on metallic base. If it is held out of perpendicular in the vertical plane, the fan pattern will become either top or bottom heavy. This will result in striping of colour coats – especially with metallic, which will show as dark and light stripes. When using direct gloss paints and clear coats, the out-of-balance fan pattern will show as either runs or dry-gloss-dry stripes. Figures 7.12, 7.13 and 7.14 show how the spray gun should be held in the vertical plane following the shape of the panel. Whilst Figure 7.15 shows how the gun should be held perpendicular to the panel in the horizontal plane.

7.10 OVERLAP

The amount of overlap required varies greatly with the type of paint material being used. It could be from 30% to 75% overlap, but generally speaking, around 40%–50% is desirable. The reason for such overlaps is to make up for the drier spray around the

Figure 7.12 Spray gun is angled slightly downwards in order to stay perpendicular to the door.

Figure 7.13 At the centre part of the door the spray gun is held horizontaly.

Figure 7.14 Here the spray gun should actually be angled further back in order to match the profile of the door.

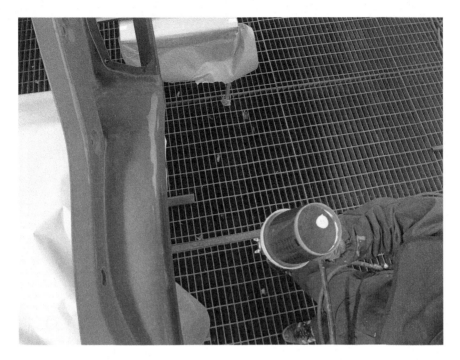

Figure 7.15 Here the spray gun is held at right angles with the panel, which is the ideal for applying consistent coatings.

edges of the fan pattern. If overlaps drop too far, colour coats will not cover evenly, and gloss coats will have dry stripes, while too heavy overlaps will result in flooding or running of base coats and running of gloss coats. It is therefore essential to get the balance right with overlaps.

Unfortunately, like so many other components of refinishing, this is largely down to experience. However, a good start is to remember that overlaps *are* important and should always be considered, especially when applying top coats. With the correct overlaps, clear coats will have an even and smooth gloss, whereas base coats will have consistent colour, with no dark or light stripes.

7.11 START AND STOP, WITH A 'FLICK OF THE WRIST'

A common mistake that many trainees make is trying to start painting in the middle of the panel, while pointing the gun at the panel. While it is possible to start painting in the middle of a panel, the gun must be angled along the surface (Figure 7.16), with the trigger depressed to the first stop to allow only air to escape, then moved into position through an arc. As the gun moves to around 120°, the trigger should be pulled a bit further to allow some paint to start coming out. As the gun is moved closer to perpendicular, the trigger should be pulled further, so that maximum paint flow is achieved as the gun reaches 90° to the panel surface.

Figure 7.16 When starting to paint in the middle of a panel, it is necessary to hold the spray gun at an obtuse angle to the panel.

Figure 7.17 When blending base coat, the masking can be used to judge the gradual fading of the colour. Inspect the blue on the masking in this picture above the graduation of full saturation of blue to no blue can be clearly seen.

This process is the beginning of blending colours (see Figure 7.17) and is a very important skill to learn, as it will prevent the nasty 'splodge' that is often the result of fully pulling the trigger while the gun is stationary and pointed directly at the panel.

7.12 LOOK AT WHAT YOU ARE DOING

It sounds obvious to watch one's paint as it lands on the panel, but many technicians forget just how important this is. Figure 7.18 shows the kind of stance a painter will pull if they are really watching what they are doing by moving their head closer to the panel. Refinishers who watch the paint and pay attention to how it is flowing out while they are applying it are able to make the subtle adjustments required. This is primarily achieved by adjusting the gun speed, but they can also observe and adjust the distance and overlap required in order to gain a consistent and smooth finish.

7.13 WORK IN PLANNED SECTIONS

When painting multiple panels on a vehicle, it is important for the technician to plan their route around the vehicle. This may be easier to do if one of the panels on the vehicle is removed due to damage, and its replacement is on a panel stand.

The resultant gap will allow the refinisher an area for overspray. Without this gap, there is a danger that one panel may have partially dried before the adjacent panel is painted, and this results in a large quantity of dry spray landing on the semi-cured paint. Figures 7.19 and 7.20 show a vehicle that required most of the

Figure 7.18 Pay attention to how the paint is going on.

Figure 7.19 This vehicle requires every panel apart from the roof and boot lid painting. The same path was used for the base coat and clear coat, starting with the driver's side corner of the rear bumper.

Figure 7.20 The technician came around the car in a clockwise direction painting the passenger quarter panel, door and wing. The front bonnet was painted next, and then down through the bumper, then the driver's side wing, door and rear quarter.

panels re-painting it was only the roof and boot lid that were left masked so a route was planned and is given in the captions for the two pictures. Although this is generally only a problem on complete repaints, it can cause problems when there is a combination of many panels, such as two front doors, both front wings, and the bonnet.

In this case, it would be recommended to start on one door work through the neighbouring wing, across the bonnet, down the other wing and onto the last door. This kind of planning is much more important for two-pack gloss coats such as clear coat or direct gloss colour coats.

Obviously, the base coats can be applied to this example front end in a different order. Assuming the doors are still fitted for a colour blend, it would be a case of applying full colour to one front wing, then allowing it to be blended into the door with a flick of the wrist.

Start at the bottom of the wing and door, then work upward before moving across the bonnet, then down the other wing, again allowing the door to be blended with an extended flick of the wrist.

While planning the route, it is recommended to work panel by panel with base coats. This of course depends on the size of the panel. A front wing is usually manageable in one sweep from front to rear, whereas a rear quarter would usually be sprayed in two, or even three smaller sections.

The panel that usually causes the most difficulty at the base coat stage is the bonnet. A bonnet should be painted from front to rear, working horizontally from one side to the other. This can get fairly uncomfortable while leaning into the centre and concentrating not to lean on the wings.

As tempting as it is to start and stop in the panel gaps, this will result in a buildup of paint on the edges of the panels, which usually results in a dark stripe on metallic base coats, or a raised buildup or run on the very edge of a freshly clear coated panel. In order to prevent this, it is necessary to divide panels into those small manageable sections as stated earlier.

Let us look, for example, at the painting of the front end described above, where the application of the base coat has already been discussed. The clear coat would be applied in an entirely different order: start with the back 90% of one front door, then the front 10% along with the section of wing behind the front wheel. The next section would be the rest of the front wing, moving up onto the bonnet.

As with the base coat, the clear coat needs to be applied to the bonnet front to rear and rear to front from one side, right across the bonnet into the top of the opposite wing. At this point, it is a case of working in reverse of the first side, clear coating the majority of the wing followed by the rear section and the front 10% of the door and finally the last 90% of the door.

A similar technique of applying clear coat in smaller sections can be described on the side of a five-door vehicle; in this instance, it is assumed that the front wing and rear quarter are the blend panels.

- Look at the side of the vehicle and divide the four panels into five or six sections.
- Start at either the front or rear of the vehicle, and spray roughly two-thirds of the panel, with clear coat flicking out in the centre of the panel, as described above.

- Once that is coated from the top to the bottom, pick up from the two-thirds point and start spraying again with that flick in spray through to roughly half way across the first door, and flick out again all the way down the door.
- Repeat this step from the first door through the shut line and roughly half way through the second door working top to bottom on one section, then bottom to top on the next.

By working along the side in the above fashion, it is possible to obtain a clean consistent finish, with minimum risk of buildup on the edges of panels. Figure 7.21 gives an example of roughly where panels should be divided up on a vehicle with two long panels. It is imperative, though, to apply with good overlaps in the joints between sections of clear coat, and it's advisable to work top to bottom on one section, then bottom to top on the next, to prevent creating extra movements for the painter.

7.14 MOVE AROUND THE VEHICLE

As has been suggested in Section 7.13, it is important to move around the vehicle properly. The author has witnessed many jobs, which could have been excellent, only to see them being ruined by a lack of finish at the bottoms of the doors and sills. As well as potentially resulting in complaints from the customer and costly rework, this kind of mistake can also leave the substrate vulnerable to water damage and result in corrosion.

A technician should stand comfortably, preferably with feet shoulder-width apart to promote balance as in Figure 7.22. Move the spray gun in a confident, but smooth, fluid motion, taking into consideration the contours of the panels being painted.

Figure 7.21 The red dotted lines are the suggested places to divide up panels when applying clear coat or direct gloss. Try as far as possible to avoid stopping and starting in panel gaps as this promotes runs on the edge of the panel.

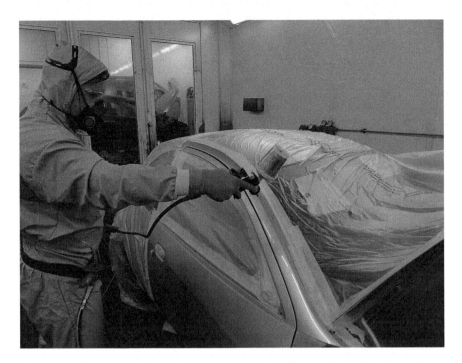

Figure 7.22 Start comfortably with arms relaxed and springy stance. Feet shoulder width apart.

A large section of vehicle can be covered at one time if the spray gun is held at full arm's length, then brought across the body, rotating the wrist to spray around corners. Of course if the refinisher prefers to start with their arm across their body as in Figure 7.23 then continue through the panel stretching to finish as in Figure 7.24 then this is fine as the next pass with the gun will be in reverse anyway.

As the refinisher works their way down the panel, they should bend gradually at the waist, and then bend their knees (Figure 7.25), in order to keep the finish consistent. As in Figure 7.26 it is often necessary to lie on the floor of the booth, in order to make sure the sills, under the bumpers and bumper recesses, are adequately covered, including underneath.

Obviously, health and safety should be considered at this point, and the floor should have no contaminants that may harm the refinisher. It is a good idea to choose overalls with knee pads for this very reason, as repeatedly kneeling down in such a fashion can have a lasting negative effect on the technician's joints.

7.15 DIRECT GLOSS

Direct gloss paints usually possess very good opacity and, as a result, can often adequately cover in two coats. However, if the primer repairs show through, more coats will obviously need to be applied. When applying direct gloss, it should be remembered that it is very much like clear coat with pigment; therefore, the application techniques are very similar.

Figure 7.23 Starting with the arm with the spray gun across ones body for maximum reach.

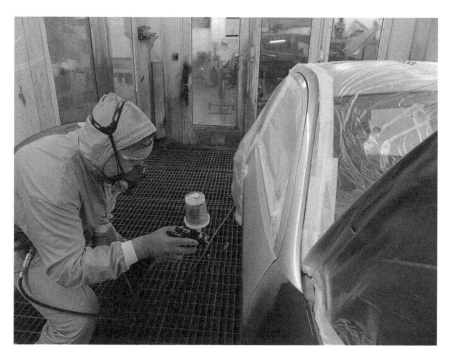

Figure 7.24 Move weight to other foot and bring arm across to maximise reach.

Figure 7.25 Bend knees while working down the panel.

Figure 7.26 It is often necessary to lay on the floor in order to ensure a good finish on sills.

MS paints are not recommended for repair procedures, for the simple reason that they are not compliant with European regulations regarding the content of VOCs. If an MS paint is to be used, for restoration processes for example, it is typically applied as two full wet coats, although the author recommends a light coat first in order to create a 'grip' coat.

Due to the amount of solvent (thinners) in MS, it will sink back further as the solvent evaporates. This means that the finish of the first full coat influences the final finish of the last coat. That's why MS should be applied as two *full* coats. Also, due to the first coat being a flow coat, MS should have a long flash-off before the final coat. Flash-off, of around 10–15 minutes, allows a lot of the solvent to escape fully. This helps to prevent solvent 'pop', which is where the remaining solvent tries to evaporate from the first coat through the top coat, which is already starting to cure.

Solvent Pop appears as tiny craters in the surface of the finish. For quick, convenient application of direct gloss, HS and UHS two-pack is recommended.

The vehicles featured in Figures 7.27, 7.28 and 7.29 were both finished with HS direct gloss. The nova in Figure 7.28 was refinished in direct gloss after the base and clear that was mixed for the re-paint left a disappointing depth of colour with too little chroma. The direct gloss gave a far superior clarity of colour. As with HS and UHS clear coats, direct gloss is usually applied with a light 'grip' coat followed by a full wet coat. Whether applying MS, HS, or UHS, in most cases the paint should be applied at an air pressure of 2 bar, but as always, the paint manufacturer's technical data should be referred to. When applying direct gloss to multiple panels, follow the recommendations in Section 7.13 for clear coats. Direct gloss is applied in the same way as clear coats, as it is essentially a clear coat with pigments added. For further details of the differences between MS, HS and UHS materials, please refer to the clear coat Section 7.17.

Figure 7.27 Vehicle in progress in direct gloss.

Figure 7.28 Direct gloss covers very well and has an excellent clarity of colour due to the pigment being right through the coats.

Figure 7.29 With the right pigments, direct gloss can give a very pleasing finish.

Before applying any paint, it should always be thoroughly stirred. Above is a sequence of images which show Figure 7.30 direct gloss paint that has settled in the tin, showing the binder floating at the top, and the pigment sunk to the bottom. Figures 7.31 and 7.32 show the pigment being stirred through the binder: it is very important to keep stirring until the material has a consistent colour and to continue for a short while to ensure all of the pigments are dispersed through the binder. Figure 7.33 shows the final colour, which is now ready to be mixed with hardener.

When stirring the paint, remember to scrape the end of the stick across the bottom of the tin, in order to incorporate the heaviest materials, which will have sunk to the bottom.

It was mentioned earlier in this chapter that direct gloss is usually applied edge-to-edge – that means the entire repair panel is painted with the colour. However, it is possible to perform a fade-out with direct gloss, and this can be performed in one of two ways:

The first (and easier) way is to take two spray guns into the booth: one with the direct gloss colour and the other with clear coat. It is imperative to make sure the two materials are compatible for this process to work correctly. The colour coat should be applied over the primer, then flicked out at the ends of the repair, well away from the edge of the panel, then immediately swap to the clear coat gun. Now apply the clear coat from the edge of the panel towards the colour, where it is then flicked into the wet coat of direct gloss. Once the first coat has flashed off, the process should be repeated for the second coat.

Figure 7.30

Figure 7.31

Figure 7.32

Figure 7.33

This process will not result in a glass-like finish immediately and, once cured, will require more polishing than a panel that was finished as an edge-to-edge repair.

The second way to blend direct gloss is to perform a fade-out in a similar way to a clear coat fade (often referred to as a 'blow-in' see Figure 7.34). In this instance, the colour is applied as a grip coat (Figure 7.35), then as a full coat, and then faded out at the end of the repair with a flick of the wrist as can be seen in Figure 7.36. While this is still wet, any dry spray created by the flick of the wrist should be wetted with thinners. This wetting is usually done with an aerosol can of spot blender or fade-out thinners.

An alternative to using aerosol thinners is to use fade out thinners as seen in Figure 7.37 is a bit messy, but results in a higher quality finish. After the colour has been applied, pour out nearly all of the paint from the gun into an empty mixing cup, then add 50% fade-out thinners to the gun pot see Figure 7.38, purge the paint from the gun away from the panel and finally use the very diluted residual paint to wet the dry spray area. Once this is done, tip the over-thinned paint into the empty mixing cup, then refill the gun pot with neat fade-out thinners, and use that to wet the edge of the over-thinned paint. Figure 7.39 shows the resultant finish of a very well executed direct gloss fade-out.

Whether the repair at hand is the more usual edge-to-edge type or a fade-out, it is very important to pay attention to how the paint is actually landing and flowing out on the panel. Direct gloss, like clear coat, will continue to flow after application; therefore, it is not only acceptable but advisable to observe some slight 'peel' when examining the surface finish immediately after application. Due to the way in which the paint flows, a glass-like finish is likely to result in a run or sag. Remember that a refinisher's role is to restore a vehicle to pre-accident condition and render the repair

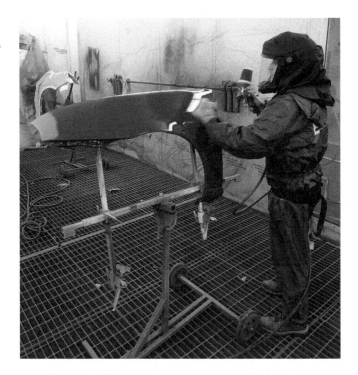

Figure 7.34 Tacking for fade-out. Note that shiny paint will not be over-coated.

Figure 7.35 The direct gloss is applied over the primer and flicked at the end.

Figure 7.36 The second coat is flicked further along the panel.

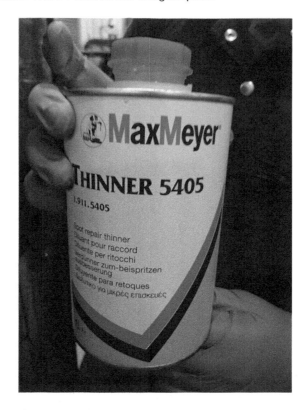

Figure 7.37 Fade-out thinner is used to wet the dry edge resultant from blending the direct gloss.

Figure 7.38 Fade-out thinner is added to the paint in order to over-thin it.

Figure 7.39 The resultant blow-in requires polishing after it has cured.

undetectable. The surface finish should replicate that of the OEM, which often has an obvious orange peel-like texture to it.

Direct gloss paint (especially red) does tend to fade far more than base and clear as can be seen in Figure 7.40. Therefore when performing an edge-to-edge repair on direct gloss paint, it is advisable to polish the adjacent panels, not only for colour matching purposes but also to restore gloss. Figure 7.41 shows a red direct gloss panel that is partway through being polished. Once finished a direct comparison can be made between the polished tailgate and the faded sides of the vehicle in Figure 7.42.

In conclusion, direct gloss provides a durable, pleasing finish at a more economical price than COB systems. The limitations are in the blending techniques, as well as the difficulties in application for metallic finishes. There is also a limitation in the types of specialist finishes such as pearls and xirallics as these are unavailable for direct gloss paint.

7.16 CLEAR OVER BASE

COB has been the accepted norm for most passenger vehicle repairs (like the repair pictured in Figure 7.43) for many years now. Base coats have the ability to offer many different finishes and hues. There is a huge range of metallic tinters that can provide subtle to very sparkly metallic finishes, as well as many colours of pearl, mica and xirallic additives.

Figure 7.40 Red direct gloss is notorious for losing gloss, fading and chalking. These are all forms of UV damage.

Figure 7.41 A machine polisher and compound easily bring back the gloss, correct hue and chroma.

Figure 7.42 Compare the gloss of the tailgate to the rest of the car. If repairs are to be made to adjacent panels, it is recommended to polish those not being repaired first.

Figure 7.43 Base coat on the door. Note the test spray to check the fan pattern on the masking.

Figure 7.44 shows a tin containing unstirred base coat, it is possible to see the components that make up its constituent parts. The sparkly parts are the aluminium particles that make it a metallic. The greenish clear liquid is the binder that is tinted in order to gain a depth of colour. If we look at the inside of the lid from the base coat tin in Figure 7.45, it is possible to see the more volatile components that condense on the inside of the lid. These will include reducers and wetting agents used to help each pass of paint melt back into the previous one. The clear residue is likely to be the wetting agent which helps with preventing striping and achieving a consistent smooth coat of paint. This should be scraped into the tin prior to stirring the contents.

Manufacturers sometimes also alter the opacity of colours. This can make them difficult to apply and gain good coverage over primers and ground coats. However, the reason why manufacturers often factor in a certain degree of translucency is that it has the ability to increase the depth of colour and metallic effect. The viewer observes not only the final top coat but also the two, or even three below it.

In the case of a base coat with poor opacity, it is very important to use the recommended ground coat. Depending on the paint manufacturer, this may be a grey shade or a non-metallic base coat of very similar hue to the final colour as this is more easily covered.

Many technicians opt to use a wet-on-wet primer of appropriate shade applied over the new or repaired panel. This is then blended into the adjacent panels with the use of fade-out thinners at the end of the blend area. The wet-on-wet primer provides a uniform finish for the first application of base coat, and more importantly, it provides

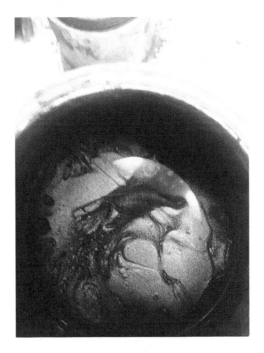

Figure 7.44 Base coat in the can showing pigments and binder.

Figure 7.45 Base coat in the lid showing pigment and wetting agent.

a ground coat of a known value shade, which helps those base coats with poor opacity to cover more easily.

An alternative to the above is to use a high-build primer with the correct value. The one in Figure 7.46 above is a very dark grey, which can be easily covered by the metallic black base used to match the car.

There are two main types of base coat available to refinishers: solvent-based and waterborne. It is recommended, as far as possible, to use waterborne base coats, as they offer lower VOCs and other environmental advantages. Pictured in Figure 7.47 is a water-based mixing scheme made up of separate tinters which can make up most vehicle colours required.

Solvent-based base coat is an older technology, but it does still have some advantages that have already been discussed in this chapter. Waterborne materials can be grouped according to the way in which their binders dry. There are several different types of binder, including microgels and latex types.

Microgels offer very clear binder properties for special-effect pigments, whereas latex binders confusingly contain no latex but are named so because, as they dry, the water evaporates first, followed by a coalescing solvent. As this solvent evaporates, it softens the binder components which then fuse together creating an irreversible film.

Figure 7.46 Value shade primer aids colour coat coverage. This saves on both material and labour costs.

Figure 7.47 These water-based tinters can be used to create most vehicle colours.

This obviously gives a very stable and strong film, even before it is overpainted with clear coat.

All base coats dry through evaporation of the solvent used in their manufacture, with solvent-based using a hydrocarbon-sourced material and waterborne base coat obviously using water as the reducer. Both types contain a binder; this suspends the pigment in the wet state and makes up the body of the colour coats in the dry state. Although base coat can be activated with a two-pack hardener, it is not recommended, and it should be treated as single-pack.

When applying a base coat, it should be remembered that it contains a lot of solvents and has a lower viscosity than other paints. Always refer to the technical data provided by the paint manufacturer. As a general rule though, solvent base can be applied as a full wet coat at a pressure of 2 bar, followed by a 5 minute flash-off, then a three-quarter coat, again followed by a 5 minute flash-off.

Solvent-based base coats usually do not require a drop coat; however, if it is showing any signs of flooding (patchiness), a drop coat should be applied at 1.5 bar. Solvent base coat has a drawback, in that the strong solvent can sometimes reactivate previous coats that have been exposed by rub throughs, so some refinishers choose to apply a very light coat first, which will dry quickly, thus helping to isolate the exposed layers.

Obviously, the *correct* thing to do is to use a rub-through primer to isolate any issues such as this. Any rub throughs are likely to cause wrinkling, like the defect in Figure 7.48. This can be flatted with P800 wet and dry to smooth it, and then it will require either a primer or a stopper to level it out.

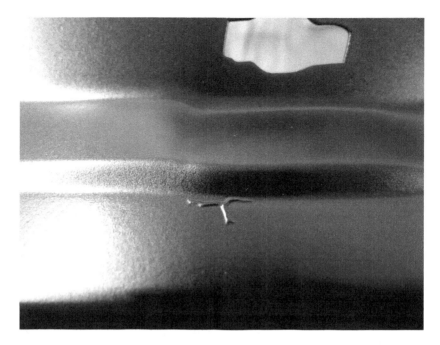

Figure 7.48 This base coat has wrinkled. This is a disadvantage of using solvent base coat. If it is applied too heavily over rub-throughs it re-activates previous paint causing the wrinkles.

Water-based paints are generally applied at 2 bar, but the technical data should be consulted too, as this will likely show one of two different ways of applying the base. Many manufacturers specify a light coat to be applied first; this is known as a 'check' coat an example of which can be seen in Figure 7.49. Its purpose is to ensure that there are no contaminants or missed preparation on the panel. It is applied lightly in order to dry rapidly; then if further sanding or cleaning is required, it can be undertaken quickly, so as not to hold up job times.

This technique was developed as the industry was moving from using solvent base as the standard to using water-based base coat. As solvent-based dries much faster than waterborne, if any defects were seen after the first full coat of solvent-based, it could be dried rapidly, then reworked without affecting job times detrimentally.

Assuming the check coat has not shown up any imperfections, a full coat of water-borne base coat is applied and dried either using booth-mounted air jets or, more commonly, handheld air driers. This wet coat should provide nearly full opacity and possess a smooth consistent film surface. However these wet coats often dry with a degree of flooding as can be seen in Figure 7.50. Once dry, a three-quarter coat is applied at 2 bar, which should provide complete opacity. Obviously, this is dependent on effective ground coating.

The three-quarter coat is applied with a slightly faster gun speed than the first coat to help prevent flooding of the metallic particles. This coat is flicked out past the ends of the first coat, to graduate any colour differences and provide the blend. After application, this coat is also dried using increased airflow.

The final coat is referred to as a 'drop' coat, and is intended to prevent the patchy appearance that can sometimes affect metallic paints that dry more slowly.

Figure 7.49 Light coat of base coat to act as check coat.

Figure 7.50 When a base coat is applied as a wet coat, it will dry patchy. This base coat is still wet, but the patchiness is obvious. When it does dry, this will still show.

Match-effect coats, often referred to as drop coats, are usually applied at a reduced air pressure of 1.5 bar. The distance between the gun and the panel is greatly increased to around 30–40 cm, and gun speed is slowed to roughly half to three-quarters of the speed of normal application.

The increase in distance allows the atomised paint to dry slightly, which prevents the flooding that can occur in wet metallic coats. This happens when the metallic particles float together in the wet paint film and gather as spots, stripes, or blotches; the light match-effect coat prevents this by being slightly drier and much thinner. Figure 7.51 shows a car that has a nice consistent finish as a result of the match-effect coat.

An alternative method of application fits very well with the use of wet-on-wet primer as a ground coat. Assuming the wet-on-wet primer is applied correctly, is nice and smooth, has no dirt nibs present, and is blended into any adjacent panels, there is no need for the initial check coat. This means a single full closed coat is applied first, which should achieve full opacity, assuming the wet-on-wet primer with the best value for the colour will have been chosen.

The full closed coat is applied at 2 bar, flicking the gun out at the ends of the blend, past the wet-on-wet primer. Once it is applied, it is immediately coated with a match-effect coat directly onto the still wet full coat. Depending on the manufacturer, the effect coat is usually also applied at 2 bar, although some suggest 1.5 bar.

This technique of application obviously has great advantages in the speed of applying colour; however, it should be noted that only certain manufacturers make base coat that can be applied in this way. Finally, do not apply base coats that should be dried between coats in this way, as they will suffer badly from flooding.

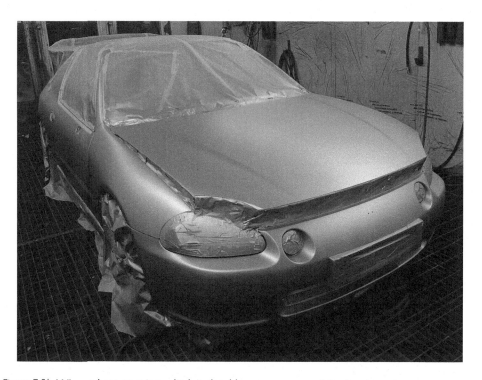

Figure 7.51 When a base coat is applied, it should appear as a consistent even coat.

While quoting the air pressure in this book, it should be noted that all pressures suggested are at the air inlet, measured with either an inbuilt pressure gauge (Figure 7.52) or a separate screw on gauge (Figure 7.53), but many paint manufacturers also quote pressures at the air cap, referred to as 'atomisation pressure'. This is a far more accurate measurement for most guns; 2 bar inlet equates to around 0.7 bar at the air cap.

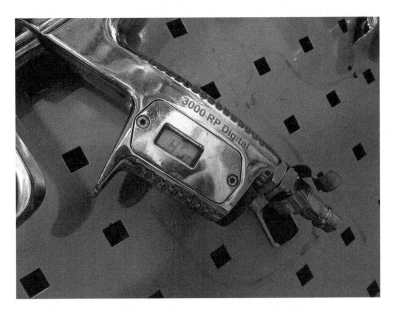

Figure 7.52 Digital pressure gauge is less cumbersome as it is integral to the gun body.

Figure 7.53 Analogue air pressure gauge is cheaper and if it breaks can be changed more easily than the integral digital type.

However, it requires specialist pressure measurement equipment which is not readily available. This replaces the normal air cap for a test spray to check pressure, and then the normal air cap is refitted for painting purposes. This information is of most importance when considering the application of basecoats, as they are influenced more than any other material by atomising air pressure. Indeed, it is the only product that requires two different air pressures to complete one application.

7.17 CLEAR COAT

Once the base coat is applied and the technician is happy with colour and blend, it should be allowed to dry fully while the clear coat is being mixed. Some technicians opt to bake the base coat, but this can increase the panel temperature too much and may prevent the clear coat from flowing out properly (although it does mean the base coat is thoroughly dried). The usual technique is to use small air driers on a stand directed at the panel, or the better option of booth-mounted air jets, like the ones pictured in Figure 7.54, which direct warmed air (often around 30°C–40°C) with a greater velocity at the panel.

Whichever method is used, it is vital to ensure the base coat is fully dried before the clear coat is applied. The main reason for this is that any solvent remaining in the base coat will evaporate through the clear coat as it dries, and this will result in Solvent Pop. In severe cases, it can even result in delamination of the paint system. The second reason to ensure that the base coat is fully dried is that, as pictured in Figure 7.55 the base should be cleaned with a tack rag prior to clear coat.

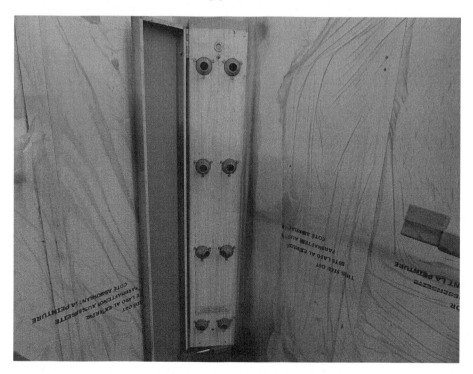

Figure 7.54 Warm air blowers are used for drying base coat.

Figure 7.55 **Some refinishers do not tack their blends. As can be seen in this picture, there are many loose particles at the end of a blend. These can be simply wiped off with a tack rag.**

If clear coat is applied over base coat which has not been tacked, there is a risk of suffering with the paint defect known as 'dissolution', which is where the metallic particles float into the wet clear coat film, and appear as extra sparkling components. This can alter face and side tones, or even if dissolution does not occur, the metallic particles in the blend can appear too bright.

When mixing clear coat, ensure that mixing cups and sticks are pristinely clean. Disposable mixing sticks are excellent for this, ensuring that there is no risk of contaminating the clear coat, as well as safer to use with disposable pots.

Most clear coats have a mixing ratio of 2:1; although some do vary, some may have a ratio of 5:3:1! As with all mixing ratios, the main constituent is the first number. If it is a two-pack material such as clear coat, the second number is the hardener, so in this instance two parts clear coat and one part hardener. Thinner may be added to clear coat to reduce viscosity and help it to flow better over the panels. Indeed some manufacturers specify that it should be used, and in those cases that instruction should definitely be followed.

When a refinisher decides to add some thinner without consulting the technical data sheet (TDS) (even if it is limited to only a few percent), it is not recommended, due to the addition of thinners increasing the VOCs in the clear coat and the likelihood of running the clear coat and gaining Solvent Pop.

Clear coats are available in three formats – the first, MS, is the oldest technology and is superseded by the other two types, HS and UHS.

MS paints should be avoided if at all possible as they are noncompliant due to their high levels of VOCs. The large particles found in MS are represented by the glass of conkers in Figure 7.56 above. As can be seen, the glass is full to the top, just like

This picture is purely for illustrative purposes and is definitely not to any scale.

MS paints have large particles like this glass of conkers.

HS paints have smaller particles like this glass of marbles.

UHS paints have even smaller particles like this glass of peas.

Figure 7.56 Example of how larger particle sizes allow more VOCs to fit in the gaps left between the particles. Medium solids have the largest particles and therefore the largest gaps between them.

the other two glasses. However, there are large gaps between the conkers (particles). These voids are filled with the VOCs in an MS clear coat.

There is a lot of VOCs in MS clear coat, and as it cures, it loses much of its film build due to the VOCs evaporating. It should therefore be applied in two closed (full) coats at 2 bar, a flash-off of around 10 minutes is necessary due to the wetness of the first coat and the quantity of solvent in it.

HS paints are usually applied as a very light coat known as a grip coat or open coat. They have smaller particles, as represented by the glass of marbles Figure 7.56. The glass is still full, but the smaller particles leave less room for VOCs in the voids, as they fit tighter together. This helps the HS clear to meet the compliancy requirements through having a lower level of VOC.

The smaller particles mean that the clear coat is denser, giving a higher film build; therefore, less material needs to be used in order to achieve the same film build as MS clears. This is the reason for being able to apply the clear coat as one open coat and one full coat. As well as being more economical due to using less material, the process is also faster, as there is a much shorter flash-off time between the two coats. Indeed, most manufacturers quote a 'single visit' technique of applying the open coat, then immediately applying the full closed coat.

In reality, when refinishing multiple panels, by the time the last one has been given an open coat, the first one should have flashed off and be ready for the final coat.

Finally, there is UHS clear coat, and this is represented by the glass of peas in Figure 7.56. As can be seen, there are very small voids between the particles, allowing even less room for VOCs. This would at first suggest that the film build is increased further – and indeed it is; however, most manufacturers still instruct refinishers to apply an open coat followed by the final flow coat.

Some other manufacturers suggest that a single flow coat is possible with no open or mist coat underneath, but in the author's experience, although this is possible, when the clear coat cures, it often drops back, producing a slightly satin finish.

There is an anomaly with the name for clear coat, that is, in the refinishing trade, it is universally referred to as 'lacquer'. This is an incorrect term, as lacquer paints are single-pack materials, like the aerosol of lacquer pictured in Figure 7.57, that dry through evaporation, whereas most clear coats are two-pack materials that dry through polymerisation after being activated by a hardener.

When applying clear coats, the final finish should always be considered, and the technical information should be referenced before mixing the material. A viscosity check is recommended. A viscosity check should be performed as pictured in Figure 7.58. Most clear coats have a suggested viscosity of around 18–21 s for a DIN4 cup. This can be adjusted with thinners if necessary, but as has already been discussed, this should be limited to a very few percent.

The air pressure at the gun inlet is usually 2–2.2 bar, but this can vary from manufacturer to manufacturer of both the clear coat and the spray gun. The correct pressure should be set, and then a test spray performed on masking sheet.

Figure 7.57 Clear lacquer does exist. But it is single pack, so it is very different from the two-pack clear coat used in most refinishing repairs.

Figure 7.58 Obviously, this is not clear coat, but coloured paint shows better in photographs, and the procedure for viscosity measuring is the same.

Once the panels are tacked to remove any loose dry spray from the base coat, application can take place. At this point, there are two camps that technicians fall into:

- The first camp argues that the edges of the panel should be coated first with a full coat, where the overspray landing on the panel is very similar to the mist coat, which is applied next, over the face of each panel. Once this is applied, it can be immediately over-coated with the flow coat.
- The other camp insists that the edges of the panels be coated last, and the faces of all panels have the open coat applied, followed by the flow coat and finally the edges.

Those who coat the edges first insist that the other technique result in overspray landing in the freshly applied film, and by coating the edges first, they are simply adding to the grip coat. Those who leave the edges till last insist that the overspray landing on the face first influence the final finish, as the last coat is applied over the offspray, and by coating the edges last, they are simply letting overspray land in the wet film, where it melts in and flows out.

In reality, refinishing technicians should try both techniques, and then decide which works best for them. While experimenting with the application of clear on the edges, technicians reading this may wish to also try holding the gun very close to the panel while increasing the gun speed twofold. Overlaps will also need to be increased with this technique.

On the subject of overlaps, clear coats should be applied with an overlap of around 50%. The last coat should be studied closely as it is applied so that the technician can vary distance, gun speed, or overlap as necessary to gain the best possible finish like the one shown in Figure 7.59.

When applying clear coat, most panels will be coated edge-to-edge; however, clear coat fade-outs (often referred to as 'blow-ins') are a very important skill and technique to get right. Being able to *not* paint to the end of the panel, while still maintaining a pleasing appearance, can save labour and material costs.

Clear coat fade-outs are a real divisive subject for refinishers, and many have their own technique which they insist is the only way to really perform a blow- in correctly. Clear coat fade-outs should be performed in a similar way as those on direct gloss two-pack paints, which appeared in images in Section7.15.

Clear coat is applied as a grip coat and then as a full coat, then graduated at the end of the repair with a flick of the wrist. The dry spray created by the flick of the wrist should be immediately wetted with thinners, which is usually done with an aerosol can of spot blender, or it can also be performed with fade-out thinners in a spray gun.

As discussed in Section 7.2, there is an alternative – which is a little tricky – but results in a very nice finish, that is, to pour out nearly all of the clear from the gun once the main area is coated, and then add 50% fade-out thinners to the gun pot. Purge the clear coat from the gun, away from the panel, and then use the very diluted clear to wet the dry spray area. Once this is done, empty the over-thinned clear from the pot, then refill the gun pot with neat fade-out thinners and finally use this to wet the edge of the over-thinned paint.

Figure 7.59 Well-applied clear coat. It should have a smooth consistent shiny finish.

Whichever technique is chosen, it is important to choose the location for the fade-out carefully. Although it is possible to perform a fade-out on just about any part of any panel, it is usual to choose either a very narrow section such as a pillar, dog leg, cantrail, or between a moulding and the edge of a door. The alternative is to perform the blow-in at the point where a panel changes direction, such as on the corner of a bumper.

Many refinishers will create what is often referred to as a 'parachute', made from masking sheet and tape at the point they want to perform the blow-in. The clear coat is flicked out to end, just before the parachute, and once the clear coat is applied and the refinisher is satisfied with the finish, the masking is torn back to reveal the shiny paint underneath, and the fade-out thinners are then applied at this point.

This technique keeps the blow-in area clean and free of overspray from both the base and the clear coat. If the technician is performing the blow-in on a wider panel, such as on the corner of a bonnet, there is no need to mask first.

One of the most important steps is to clean the panel thoroughly with a tack rag after the base coat is applied, and before the application of the clear coat and fade-out as described. Whichever of these scenarios is played out, there are several other important steps prior to masking and applying the clear coat:

- The first step is to polish well past the blow-in area, in order to make sure that there are no surface contaminants that may appear as dirt nibs or contamination in the finish where the clear coat fades out.
- Then clean this area and the rest of the panel, with solvent panel wipe to remove any polish residues.

The area that is to have clear coat applied to it should be flatted with a suitably fine abrasive paper such as P1000 or P1200, or alternatively grey abrasive fleece can be used here to provide a mechanical key. The flatted area should meet the polished area, like in Figure 7.60, where the technician plans to perform the fade-out.

The polished section, as well as being contamination free, also helps with the next step of polishing up the area where the fade-out thinners have been applied.

Again, this is a divisive opinion, as some technicians will flat the area down past the section they intend to apply the fade-out thinners to. They argue that there is then no chance of the clear coat accidentally lifting from the edge, but it does also make it very difficult to return the area to full gloss.

Once the clear coat is cured, the blow-in should be polished. This is a risky procedure, but it is necessary in order to complete the repair properly.

The area that has clear coat applied should have any dirt nibs removed through flatting, just like on a panel that has been clear coated edge-to-edge.

The area that has the fade-out thinners applied to it should not be sanded, as the clear coat at this point is very, very thin.

In order to polish up the entire panel, start at the far end, away from the clear coat, and gently machine polish back towards the clear coat. This will remove the overspray resultant from the panel being unprotected during painting.

As the technician passes over the area that has the fade-out thinners on it, they should take great care, as it is extremely easy to break the blow-in and expose the edge of the clear coat. This will appear as a wavy line on the surface, and if this does happen, the repair will have to be started again.

Figure 7.60 This wing has been prepped for a blow–in, the shiny paint will not be over-coated.

As long as that area does polish up correctly, the technician can move onto polishing the actual area of clear coat that has been de-nibbed; at this point, polishing can be undertaken in exactly the same way as a panel that has the clear coat applied edge-to-edge.

Correctly applied and polished clear coat provides a very pleasing gloss when viewed in the sun. Many coats of clear coat are required to even out rough metal flake – the roof in Figure 7.61 has six to eight coats. They were applied two at a time, and then flatted with P800 after curing and before the next two coats.

7.18 SUMMARY

The application of top coats is a highly skilled and difficult procedure, and there are many factors that can make the final finish appear incorrect. Part of being a successful and conscientious refinisher is removing those factors that may make the final finish substandard. Much of this is down to experience; however, there are many steps that should be undertaken that refinishers with many years of experience sometimes forget. These can help an inexperienced refinisher produce a better quality finish. Those key points are as follows:

- Keep everything clean, and reclean after every step. This may mean panel wiping after masking or using a tack rag to remove dry spray after base coat.
- Spend your time choosing the correct colour and finish, Match the metallic, and choose direct gloss or COB.

Figure 7.61 Custom metal flake paint with multiple coats of clear coat applied two at a time. After the two coats, the clear coat is cured and then sanded before the next two coats are applied. This evens out the extremely rough finish resultant from the application of clear coat crammed with large chunks of glitter.

- Observe the environment in which you intend to spray, and change or adjust it if necessary. This includes temperature, air quality and lighting. It is difficult to alter humidity however, which can have a massive effect on the final colour.
- Spend time setting up the spray gun. Fan pattern/size, fluid flow rate and air pressure are all very important factors.
- Be mindful of spray gun distance, angle and speed across the panel.
- Check all preparation thoroughly before and after masking.

When applying refinishing top coats, remember that the whole point is to try and replicate the original appearance; the repair should be undetectable in finish, colour and gloss.

Chapter 8

Health and safety in the refinishing workshop

As with any industry, there are health and safety issues associated with the procedures we undertake, the chemicals we use and the environment in which we work. Some of these dangers can have immediate effects, and some may take a longer time to affect individuals. There are, of course, procedures in place to prevent exposure to harm, and it should be remembered that the safest thing to do is *remove* the danger, and if that just cannot be done, appropriate personal protective equipment (PPE) and respiratory protective equipment (RPE) should be employed to reduce the risks.

If at any point you are unsure of any Health and Safety protocol, there is guidance and advice for employers and employees on the Health and Safety Executive (HSE) website (hse.gov.uk). *The HSE is a UK government agency responsible for the encouragement, regulation and enforcement of workplace health, safety and welfare, and for research into occupational risks in Great Britain.*

8.1 PPE AND RPE

PPE covers all safety equipment technicians should wear to protect themselves, apart from those items that are designed to protect their respiratory system, which are classed as RPE.

8.1.1 Head

Many refinishers will not bother with items to protect their head from injury, as our profession does not often require us to work under vehicles as with mechanics, or in areas of restricted access, as with panel beaters. However, it is a good idea to wear a 'bump' cap, like the one pictured in Figures 8.1 and 8.2, if working in the boot area or under the bonnet. Bump caps often resemble a baseball cap and are more comfortable to wear than a hard hat. Some air-fed masks have bump caps built in to add protection to refinishers while they are applying paint in restricted areas.

If a technician has hair long enough to tie back, then they should do so, as this will help stop paint products getting in their hair, or their hair getting into a wet paint film. More important though, it will help prevent accidental 'scalping', due to hair getting caught in rotating equipment such as polishers and dual action (DA) sanders. On the subject of long hair, long beards should be tucked into overalls or held back with beard nets, as some unfortunate people have had their beards tangled in rotary polishers.

Figure 8.1 Bump caps usually resemble a baseball cap; the peak offers a little protection for the eyes.

Figure 8.2 The inside of the bump cap has a hard plastic shell with some soft foam lining.

8.1.2 Eyes

It is highly recommended that paint technicians wear safety spectacles, like the ones pictured in Figure 8.3, or goggles for virtually every job that can be undertaken in the refinishing workshop. It is therefore good practice to get into the habit of wearing them at all times in the working environment.

The range of materials that can accidentally enter one's eyes in a refinishing workshop is alarming and includes wet paint or thinners when mixing paints, gun wash when cleaning spray guns, panel wipe when cleaning panels, abrasive particles and paint flakes when sanding areas for repair, metal sparks when removing corrosion, polish and foam particles when polishing panels and many other tasks too numerous to mention here.

8.1.3 Hearing

There are many noisy tasks in the refinishing workshop which require ear protection, from the use of DA sanders and polishing machines, to extraction units and air dusters. The choice for protecting one's hearing comes down to personal preference and includes two options: the first being the rather cumbersome ear defender, like the ones pictured in Figure 8.4, which is difficult to wear along with safety glasses and dust mask, or the second being ear plugs, such as the ones in Figure 8.5, which are more comfortable to wear, but have the drawback of being rather unpleasant to reinsert if they have been worn before. Either way, they are an important part of refinishers' PPE but are often overlooked.

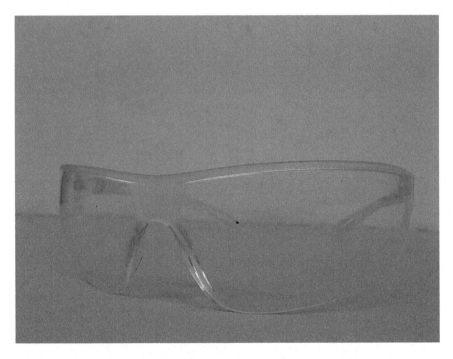

Figure 8.3 Safety glasses should be worn for many jobs in the workshop, so some technicians choose to wear them all the time in the workshop.

Figure 8.4 Ear defenders provide protection from loud noises.

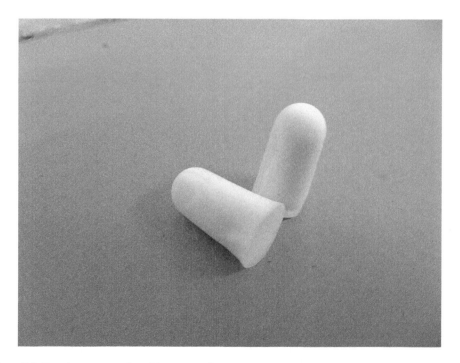

Figure 8.5 Ear plugs are comfortable to wear but are not very pleasant to reuse.

8.1.4 Hands

Dermatitis is a serious skin complaint which can be very uncomfortable, but it is easily avoided by using the correct protection. The technician who suffers with the dermatitis pictured in Figure 8.6 worked in body shop who were reluctant to buy disposable gloves the result is a skin complaint that flares up if he has any contact with panel wipe or thinners. He now observes all the following recommendations.

The first type of protection is barrier cream. This should be applied when a refinisher first enters the workshop and reapplied every time they wash or wet their hands. However, barrier cream alone is not enough to defend against exposure to solvents and paints. For general workshop tasks such as putting panels on stands, dry sanding and colour matching, it is recommended to use a glove with a nitrile palm and fingers, and a breathable fabric back (Figure 8.7) that prevents hands perspiring overly, as this can also cause dermatitis.

When using solvents and wet paints, it is recommended to use disposable nitrile gloves, such as the ones pictured in Figure 8.8. These are more resilient and less likely to cause reactions than latex gloves. They are solvent proof for short exposure, such as cleaning panels with panel wipe. However, they will let strong solvents soak through and attack the skin if exposed for extended periods of time, such as for gun cleaning. In this instance, chemical resistant gloves, such as those pictured in Figure 8.9, should be worn. These may reduce dexterity, but they are necessary in order to help prevent skin exposure to gun wash.

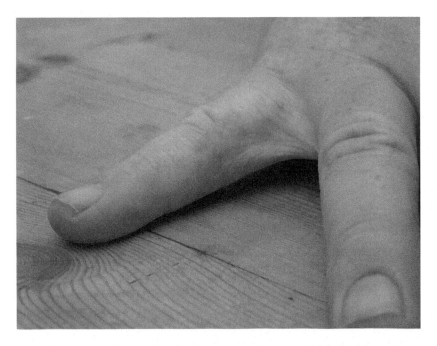

Figure 8.6 Dermatitis is usually found between the fingers where solvents can gather; this is a fairly mild case, although it brings a lot of irritation to the individual who has to constantly moisturise to prevent painful red sores.

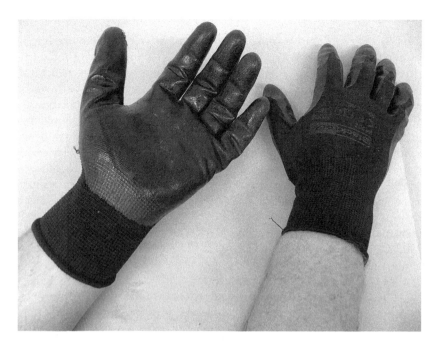

Figure 8.7 These fabric backed workshop gloves provide breathable protection from most workshop jobs. They are not suitable for using with any wet paint or thinners. The rubber palms provide grip but are not solvent resistant.

Figure 8.8 Nitrile gloves should be worn for many tasks including panel wiping.

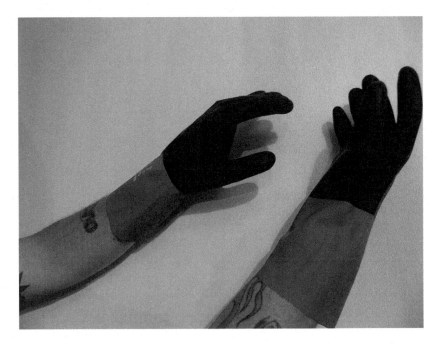

Figure 8.9 Chemical resistant gloves for cleaning spray guns.

8.1.5 Body

Although many workshops are now adopting hardwearing work trousers and polo shirts, there are many who still prefer the extra protection of overalls, as they have long sleeves and fewer gaps, and can be worn over one's normal clothes. There are flameproof types for panel beaters, as well as generic ones, which should be soft fabric like the ones pictured in figure 8.10.

As a refinisher, it is an excellent idea to keep one's clothes clean and free from contamination such as dust from sanding and polish residue, so when it is time to paint, there is less chance of contaminating the wet surface.

Spray suits are the most important type of overalls for paint technicians. These may be of the paper disposable type or the reusable soft polyester or nylon types, such as the one pictured in Figure 8.11. Whichever version is chosen, they should be kept clean and dust free, and if the reusable ones are chosen, they should be washed regularly to prevent a buildup of paint layers.

Spray suits are usually supplied with a hood, which should be utilised in order to prevent loose hair contaminating the wet paint film, as well as protecting the technician's head from atomised paint and off spray. The cuffs of both the ankles and the wrists are elasticated to prevent the ingress of atomised paint or the overalls being dragged through the wet paint film.

8.1.6 Feet

It is recommended to wear steel toe-capped boots, like the well used pair in Figure 8.12, although many technicians opt for stout boots, which are also acceptable. Bear in mind

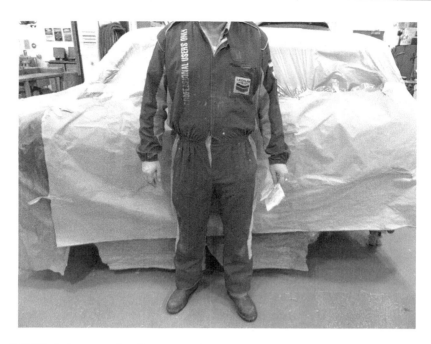

Figure 8.10 Workshop overalls offer an extra level of protection for general workshop duties and preparatory work.

Figure 8.11 Spray suits should be worn with the hood up, partly to prevent loose hairs and hair products getting into the paint, but more important to protect the refinisher from exposure to isocyanates and VOCs.

Figure 8.12 **These battered old steel toe-capped boots are very comfortable. Grip well and provide the safety of thick leather and steel toe caps.**

that there are numerous activities where feet are endangered – for instance, moving heavy replacement panels, pushing immobilised cars, or lifting the heavy grills that make up the booth floor. Soft trainers or dress shoes are not recommended, as they do not possess the chemical resistant soles of dedicated work boots, which should still offer some modicum of grip if thinners or gun wash has been spilt on the floor.

8.2 RESPIRATORY PROTECTIVE EQUIPMENT

There are many unpleasant chemicals involved in the refinishing trade that may damage your health through breathing them in. These may be dust, spray mist, or vapors. The various types of RPE are outlined below.

8.2.1 Disposable dust masks

Pictured in Figure 8.13 is a typical paper disposable mask, these are only really suitable for low-volume dust areas, such as when dry – sanding while using good extraction. They offer limited protection and are definitely *not* suitable for wearing while spraying.

8.2.2 Reusable dust masks

These usually have a soft rubber face cup that fits snuggly to the face, but they are not suitable for people with beards, as the seal is compromised – unless they smear their

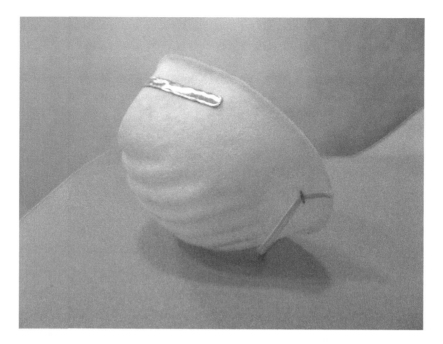

Figure 8.13 Disposable dust masks offer little protection but should be worn when sanding using extraction.

beards with petroleum jelly! These masks have a variety of ratings, some of which will even filter out asbestos.

Depending on the manufacturer, these masks will have either bonded-in filters, like the one to the right of the picture in Figure 8.14, or separate filters, like the one pictured to the left in Figure 8.14, that can be changed after a given amount of time or degree of contamination. With the correct rating, these can be used for spraying single-pack paints such as anti-stone chip, but the manufacturer's technical details should first be consulted. Filter masks should not be used for two-pack paints as the isocyanates they contain cannot be filtered out.

8.2.3 Air-fed masks

There are two types of air-fed masks: full face and half masks. Figure 8.15 shows a half mask whilst Figure 8.16 shows the more popular full face air-fed mask. It is recommended that the full face mask is used as often as possible, but the half mask may be worn if the refinisher prefers their comfort and visibility. It should be remembered that if the refinisher has more than a day's stubble, the half mask will not seal properly, and goggles and a hood should also be worn with this type of mask.

Whether the mask is half or full face, the air supply must be clean and tested regularly to ensure it is suitable, doesn't contain any oil mist, and isn't contaminated by the foul air from the spray booth. It is important to ensure the hood of the air-fed full face mask is used to prevent foul air leaking in around the refinisher's head.

Figure 8.14 These reusable masks are quite different to wear. The red one provides very thorough filtering, but the weight of the replaceable filters can make it heavy to wear. The blue one is much lighter and more comfortable, but the filtration is not as thorough. Both are only really suitable for dry sanding and applying some single- pack paints.

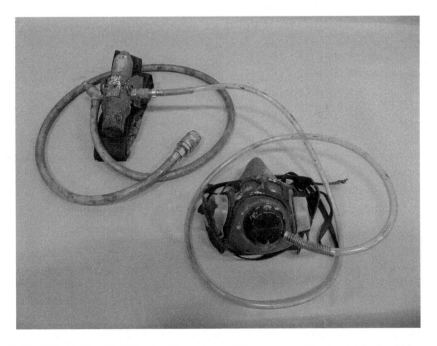

Figure 8.15 Air-fed half mask. This should not be used by anyone with a beard. It should be used in conjunction with safety glasses and the hood of the spray suit raised.

Figure 8.16 Air-fed full face mask can be used with the same belt pack. The visor should have a removable clear sheet that is changed when obscured with overspray. There should be an audible or visual warning system to warn the technician when there is not enough air being supplied to the mask.

One of the most important things to remember is not to lift the visor after spraying has taken place for the length of time the booth takes to clear.

Whenever paints containing isocyanates are sprayed, an air-fed mask must be worn. Full face air-fed masks must be fitted with either an audible or a visual warning to let the technician know if the air supply drops to a dangerously low level, while half masks will not provide enough air for a painter if the pressure drops, so they will need to get out of the booth and remove the mask as soon as possible.

Full face masks will have some type of removable film to fit over the visor, so that when they become covered in overspray and affect the technician's vision, the film can be removed and replaced with a new one.

8.3 FIRE EXTINGUISHERS

Unfortunately, accidents *do* happen, and sometimes those accidents result in a fire. The usual advice applies if one discovers a fire, and that is to raise the alarm and immediately leave by the nearest fire exit. However, if the fire is small enough to fight, then it should be. It is therefore vital that the correct type of fire extinguishers are used, as it can be very dangerous to use the wrong type. Fire extinguishers should have labels on the wall where they are stored like the ones picture in Figure 8.17, the labels tell the user what type of fire they can be used on.

All fire extinguishers nowadays are red, but the contents are indicated by a coloured label. The following table shows the colour of the label, the contents, and what type of fire the extinguisher can and cannot be used on:

Water	Dry powder	Foam	CO$_2$	Wet chemical
Limited to solid material fires. Soaks anything they are deployed upon.	May be used on most types of fire. Cause lots of mess and makes bare metal rust.	Similar to water but can be used on liquid fires as well.	Safe on electrical and flammable liquids. Should not be used in confined spaces.	Safe for solid fuel fires and cooking oil. Needs to be fully discharged.
Do use on type A fires. Do not use on type B or electrical fires.	Do use on type A, B, C, and electrical fires.	Do use on type A and B fires. Do not use on electrical fires.	Do use on type B and electrical fires. Do not use on type A and D fires.	Do use on type A and F fires

- *Types of fire*
  - Class A combustible materials (solid material fires)
  - Class B flammable liquids (paint, thinners, etc.)
  - Class C flammable gases (acetylene, liquefied petroleum gas, etc.)
  - Class D flammable metals (lithium, potassium)
  - Electrical fires (Computers, electrical equipment, etc.)
  - Class F deep fat fryers (cooking oil fires)

Figure 8.17 Fire extinguisher signs found above the fire extinguisher storage.

8.4 HEALTH AND SAFETY SIGNS

The health and safety signs dealt with in this section are not exclusive to the refinishing trade; however, they are very important, as they provide us with the information required to maintain safe practices for the people operating equipment and those around them. Safety signs are colour coded to help with identification.

1. **Prohibition sign (Figure 8.18):** The HSE says, "A sign prohibiting behaviour likely to increase or cause danger (eg 'no access for unauthorised persons')". Prohibition signs are telling you what things you shouldn't do. They are round and have a red border with a diagonal line running through a pictogram of the thing that shouldn't be done.
2. **Warning sign (Figure 8.19):** The HSE says, "A sign giving warning of a hazard or danger (eg 'Danger: Electricity')". *These* warning signs are triangular and have a yellow background with a black border and a pictogram of the thing that is dangerous in the middle. They are warning you that there is a danger present which you should be watchful of.
3. **Mandatory sign (Figure 8.20):** The HSE says, "A sign prescribing specific behaviour (eg 'Eye protection must be worn')". Mandatory signs are circular and have a blue background with a white border and a pictogram of the instruction or item of PPE to be worn. They are signs that are instructing you of something you either must do or must wear.
4. **Emergency escape or first aid sign (Figure 8.21):** The HSE says, "A sign giving information on emergency exits, first aid, or rescue facilities (eg 'Emergency exit/Escape route')". Emergency signs have a green background, white border and also the instruction as a pictogram in white. They offer advice on what to do in an emergency, how to leave, how to stop something, or where an emergency piece of equipment is.

Figure 8.18 The red signs are the prohibition signs, and the yellow sign is a warning sign.

Figure 8.19 Warning sign.

Figure 8.20 This should be Mandatory not Prohibition sign.

Figure 8.21 First aid sign.

8.5 REGULATIONS AND ACTS

There are many Acts and Codes of Practice that apply, but are not exclusive to the refinishing industry, and also apply to other industries.

8.5.1 Health and Safety at Work etc. Act 1974

The HSE says, "The Health and Safety at Work etc. Act 1974 is the primary piece of legislation covering occupational health and safety in Great Britain. The HSE, with local authorities (and other enforcing authorities) is responsible for enforcing the Act and a number of other Acts and Statutory Instruments relevant to the working environment".

It is an Act that gives guidance to both employers and employees, on what should be provided in the workplace to provide a safe and healthy place of work. It also covers customers and passers-by. The HASAWA is not just for the refinishing industry but for any industry in any location.

8.5.2 Control of Substances Hazardous to Health

Control of Substances Hazardous to Health (COSHH) is a set of regulations that aim to help employers and employees protect their health when using hazardous substances. It covers all industries including refinishing and provides guidance in the safe use of hazardous substances, including using the correct PPE.

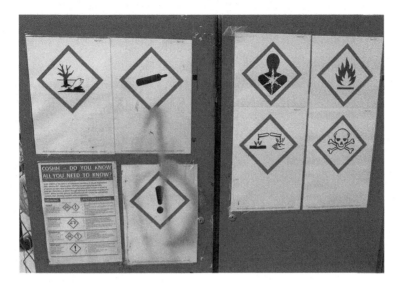

Figure 8.22 **COSHH signs.**

Below is the list of types of substances covered by COSHH. Next to each is an example of a substance found in the refinishing industry.

- Chemicals: Thinners
- Products containing chemicals: Paints
- Fumes: Vehicle exhaust
- Dusts: Paints and fillers when dry sanding
- Vapors: Styrene from open filler tins
- Mists: Atomised paint
- Nanotechnology: Not relevant to refinishing
- Gases and asphyxiating gases: Isocyanate
- Biological agents: Not relevant to refinishing
- Germs that cause diseases: Occasionally, windscreen washer bottles will develop Legionnaires' disease.

As well as the correct PPE and RPE, COSHH gives advice on the storage, extraction and filtration of dangerous chemicals. Below are the pictograms used by the HSE to warn users and handlers of the dangers inside an environment, a storage area, or a container. These are found on packaging, including household cleaning products, doors to workshops and factories and the outside of cupboards as pictured in Figure 8.22.

Hazard pictograms alert us to the presence of a hazardous chemical. The pictograms help us to know that the chemicals we are using might cause harm to people or the environment.

Explosive (Symbol: exploding bomb). When a vehicle has been involved in an accident, it will often require the passive safety devices to be replaced. This means obtaining new airbag units, seatbelt tensioners and side curtains. These components are deployed via explosive charges. When the new ones are stored in the workshop, they must be kept in a locked metal cupboard able to withstand the blast if a component is accidentally deployed.

Flammable (symbol: flame). This symbol will be found on the side of gun wash, thinners and solvent panel wipe containers among others.

Oxidising (symbol: flame over circle). The peroxide found in many body filler hardeners is an oxidising agent. So this symbol is often found on the packaging.

Corrosive (symbol: test tube pouring corrosive liquid on a hand and a surface). This symbol is found on paint stripper containers.

Acute toxicity (symbol: skull and crossbones). This symbol is also found on paint stripper containers.

Hazardous to the environment (symbol: dead tree and fish). Most paints are hazardous to the environment, so this symbol is seen on thinners and gun wash.

Health hazard/hazardous to the ozone layer (symbol: exclamation mark). This symbol is found on tins of filler and hardener.
A couple of new pictograms are also introduced:

Serious health hazard (symbol: silhouette of a person with star infront). This is found on a range of paint products, including some panel wipes, some etch primers and specialist coatings.

Gas under pressure (symbol: gas cylinder). This symbol should be displayed on the outside of the cupboard where aerosols are stored.

The COSHH regulations are much more than a set of pictograms and should be considered as a guide for the safe use, storage and disposal of the types of chemicals and other products considered to be hazardous.

8.5.3 Safety data sheets

Safety data sheets (SDS) were previously known as material safety data sheets (MSDS) and provide information on chemical products that will help users of those chemicals to make a risk assessment. They describe the hazards the chemical presents and give information on handling, storage and emergency measures in case of accidents.

When a paint material retailer supplies any refinishing products, they must also provide access to the SDS. These used to be stored in the workshop as paper-based documents in a folder. Now that virtually everyone has access to the Internet on their mobile phones as well as workshop PCs, SDS are provided online. This means that access can be gained virtually immediately to new products and any chemical ingredient changes that may have taken place with existing products.

SDS are directly linked to the COSHH regulations, as they inform both the supplier and the user how to safely store, transport and use the refinishing materials concerned.

8.5.4 PUWER – Provision and Use of Work Equipment Regulations

The following extracts have been edited from the HSE website:

Provision and Use of Work Equipment Regulations (PUWER) places duties on people and companies who own, operate, or have control over work equipment. It also places responsibilities on businesses and organisations whose employees use work equipment, whether owned by them or not.

PUWER requires that the equipment provided for use at work be

- Suitable for the intended use.
- Safe for use, maintained in a safe condition and inspected to ensure it is correctly installed and does not subsequently deteriorate.
- Used only by people who have received adequate information, instruction and training.
- Accompanied by suitable health and safety measures, such as protective devices and controls. These will normally include emergency stop devices, adequate means of isolation from sources of energy, clearly visible markings and warning devices.
- Used in accordance with specific requirements, for mobile work equipment and power presses.

Generally, any equipment that is used by an employee at work is covered, for example, hammers, knives, ladders, drilling machines, power presses, circular saws, photocopiers, lifting equipment (including lifts), dumper trucks and motor vehicles. Similarly, if you allow employees to provide their own equipment, then it will also be covered by PUWER and you will need to make sure it complies.

In short, PUWER covers most tools and equipment found in a body shop. The tools should be suitable for use and in a safe condition.

8.5.5 RIDDOR – Reporting of Injuries, Diseases and Dangerous Occurrences Regulations

Reporting of Injuries, Diseases and Dangerous Occurrences Regulations (RIDDOR) is a regulation that puts duties on employers, the self-employed and people in control of work premises to report certain serious workplace accidents, occupational diseases and specified dangerous occurrences.

There are many people who should report accidents and dangerous occurrences according to the HSE; however, those most relevant to the refinishing industry are as follows:

- Employers
- The self-employed
- Members of the public
- Those who have been injured
- Employees
- Gas suppliers
- Gas engineers

The types of injuries and diseases to be reported are as follows:

- Deaths and injuries caused by workplace accidents
- Occupational diseases
- Carcinogens, mutagens and biological agents
- Specific injuries to workers
- Dangerous occurrences
- Gas incidents

Many paint shop materials include carcinogens and mutagens, so the SDSs should be consulted to find out what materials these are. Obviously, many spray booths use gas for their heating, so gas incidents are a very relevant concern within the refinishing trade. It is important to record accidents, as this can show trends in diseases and accidents that are perhaps linked to a change of technology or working practices. These can then in turn bring forth changes in recommended working practices and PPE requirements.

8.5.6 Environmental Protection Act (EPA) 1990

The Environmental Protection Act (EPA) provides the structure of waste management in the United Kingdom. It is an Act of Parliament that covers the control of waste materials on land, in water and airborne. This affects the refinishing trades in many ways, including how spray booth filters, masking materials and abrasives are disposed of; how used gun wash is recycled; and how the foul air from the spray booth is expelled to the atmosphere.

As of 2008 the Secretary of State for the environment was able to prescribe what materials constitute controlled waste. They can also set limits on the quantities produced, as well as give guidelines for the correct way to do so. This goes as far as dictating how far above the roof line an extraction chimney should finish.

The Environment agency are the department tasked with enforcing the laws around the controlled waste, so it provides businesses and individuals with instructions on how to correct any errors with the way they are disposing of their waste. If the offending company or individual does not conform to the given instructions, they may be taken to court and fined or even, in extreme cases, imprisoned.

As well as those actually producing the waste, the EPA also imposes a duty of care on suppliers to ensure they are supplying products in a responsible manner. This means that if they suspect that the end user may not dispose of the resultant waste correctly, then they should not supply them with those materials.

The overall aim of this Act of Parliament is to reduce harmful waste in the environment and provide a form of punishment for those who contravene the rules prescribed by the Act.

8.6 WORKPLACE POLICIES

Health and safety is entirely reliant upon the people in the workplace behaving in a responsible manner. This may include the obvious points, such as reporting accidents, maintaining machinery and choosing the correct PPE, but also includes how they conduct themselves.

It is common knowledge that alcohol and recreational drugs affect peoples' inhibitions and their ability to move in a coordinated manner. Therefore, most workplaces have a policy where alcohol and recreational drugs cannot be consumed in working hours, and the technician should not be under the influence of either when at work.

This is firstly to lessen the chances of accidents caused through reckless behaviour and poor judgement, but also to make sure that the quality of work is up to an acceptable standard, and that vehicles are repaired in a manner which leaves them safe to use on the public highway.

It has been illegal to smoke in enclosed workplaces in Europe since July 2007, but even before this, smoking was banned from accident repair workshops due to the fire risks caused by solvents, filler dust and masking materials. When 'going for a smoke', employees should do so only in areas designated by the employers.

Having previously covered workwear from the point of view of PPE, it is now time to regard it from the aspect of 'presentation'. It is important to present oneself in a clean and professional manner. This means keeping overalls clean and using vehicle protective equipment (VPE) such as seat covers and floor mats. These features should help portray a professional image, which should give customers the reassurance that their vehicle is going to be repaired to an acceptable standard.

Eating and drinking should be undertaken in a designated area, for several reasons; as well as portraying a professional image by not having food laying around the workshop, eating in an area full of toxic and dangerous materials can be very dangerous. Technicians should wash their hands thoroughly before eating and should make sure that the area in which they are eating is not contaminated with the poisonous materials found around accident repair workshops.

To recap – there are many workshop policies that are put into place to help protect employees and members of the public from health and safety issues, but others are recommended simply to help the company portray a higher quality and more professional image.

Chapter 9

Tool and equipment maintenance

Having discussed the importance of knowledge, practice and skill within the refinishing industry, it is important to emphasise the significance of having good quality tools, in a suitable condition, to ensure safe reliable operation at their full capability so that the work is carried out to an adequate level of quality.

The tools used within the refinishing industry are generally expensive; however, with correct cleaning and maintenance, they will provide years of trouble-free service. Before undertaking any form of tool maintenance, it is important to make sure that you have had the correct training in order to maintain the tool in a manner which will leave it safe to use and operating correctly. It should also be noted that larger pieces of equipment may cause injury to the person undertaking the maintenance if it is not done correctly. Therefore, if you are asked to perform tool or equipment maintenance and feel that it falls outside the remit of your job, you are quite within your rights to ask for training to ensure that you know how to do it correctly.

9.1 HAND TOOLS

When checking hand tools, look for any obvious damage:

- Are there any cracks?
- Are all the components in place?
- Are any moving parts doing so smoothly and easily?
- Are electrical flexes in good condition with no splits or cuts to the insulation?
- Are there any air leaks?
- Has the tool been properly cleaned after the last use?

Refinishers don't have too many different types of hand tools to worry about maintaining;. There are basic tools such as screwdrivers and spanners for stripping components; air tools such as dual-action (DA) sanders, mini grinders and belt sanders; and electrical tools such as polishers, grinders and heat lamps.

Air tools usually have a statement embossed on them saying 'oil daily' (see Figure 9.1), and to do this, 'air tool' oil should be used. This can be obtained from tool suppliers, and as in Figure 9.2, one drop should be added to the air inlet each day. In reality, there are very few workshops where this happens, as air tools will run for months without oil; however, when they seize, they are very difficult to free up. Do not be tempted to

Figure 9.1 Air tools usually have oil daily either cast stamped or engraved on the side of the tool.

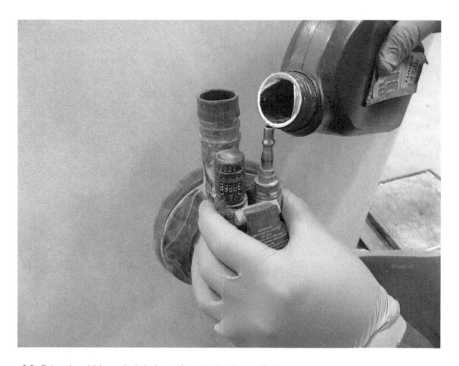

Figure 9.2 DAs should be oiled daily with a single drop of oil.

add more than one drop of oil at a time, as excess oil will be sprayed over the panels being worked on, which obviously brings the possibility of paint defects, such as silicone contamination.

Tools such as DAs should also have the backing pad checked for condition. If it is in poor condition, a new pad should be fitted – seek help with this if required. It is dangerous to use a DA in this state due to the risk of the broken piece flying off. Mini grinders and belt sanders have been designed so that the consumable parts, that is, discs or belts, can be changed easily. Air tools are used widely in the refinishing industry due to their simplicity, which makes them dependable, and they are often lighter than the electrical equivalent.

Air-fed masks should be kept in good working condition, as they are important safety equipment. Any broken parts should be replaced, and the filter found on the belt pack should be changed when out of date or if it has reached the maximum number of operating hours. Figure 9.3 shows the filter pack on the air-fed belt which has a clear screw off cover for the filter. Figure 9.4 shows the same filter pack opened up to check the condition of the filter.

There are many electrical tools that have been thrown out simply because the brushes have burnt out. These are a service item which many people do not understand how to change, and indeed some electrical tools are supplied with spare brushes and instructions on how to change them. It is usually a very simple task like in the Figures 9.5, 9.6, 9.7 and 9.8 which show the sequence of changing and checking brushes.

As has already been stated, if the electrical flex is damaged, the tool such as the badly damaged flex on the electric grinder in Figure 9.9, shouldn't be used. To replace

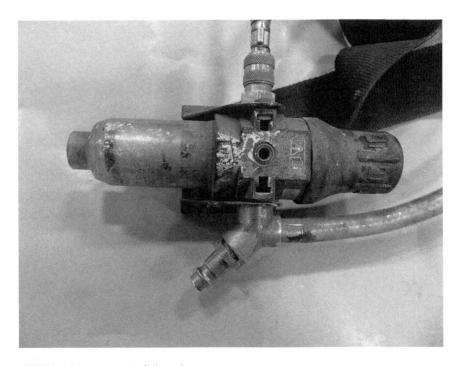

Figure 9.3 Filter housing on air-fed mask.

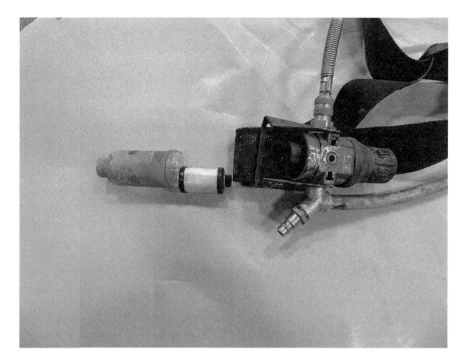

Figure 9.4 Changing filter in belt pack for air-fed mask.

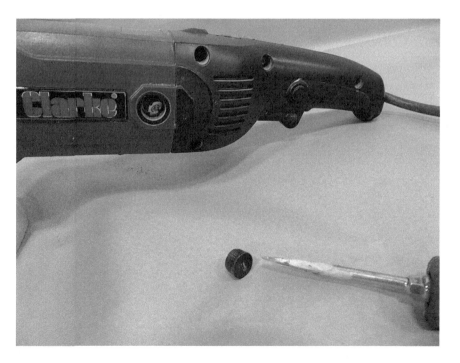

Figure 9.5 Undo the brush cap with a screwdriver. These caps are soft plastic, so choose a well-fitting screw driver. Note the end of the brush in the aperture.

Figure 9.6 Compare the new and old brushes. The older one (top) is obviously shorter, although it does not require replacing at this point so maybe re-installed.

Figure 9.7 The new brush is simply pushed in the correspondingly shaped orifice. Make sure it is fully seated.

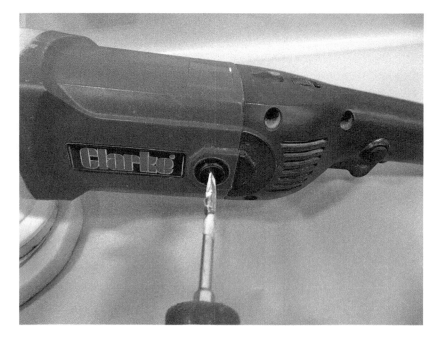

Figure 9.8 The brush cap is replaced. Make sure it is not cross threaded, and only do it up till it resists. Remember that it is plastic and is easily damaged.

Figure 9.9 This angle grinder has a badly damaged flex. This is dangerous and should be replaced before the tool is used anymore.

this is a more difficult job, and professional help should be sought before attempting to fit new electrical flex to any equipment.

All electrical equipment that can be moved such as the electric polisher featured in Figure 9.10 should be checked annually with a portable appliance test (PAT), and all tools that have passed PAT have a date label (Figure 9.11) stuck to them, often wrapped around the flex.

Many people do not realise that PAT is not a legal requirement but a strongly recommended procedure. The Health and Safety Executive says, "The Electricity at Work Regulations 1989 require that any electrical equipment that has the potential to cause injury is maintained in a safe condition. However, the Regulations do not specify what needs to be done, by whom or how frequently." (They don't make inspection or testing of electrical appliances a legal requirement, nor do they make it a legal requirement to undertake this annually.) Larger equipment such as vehicle lifts and spray booths do not require PAT, as they are obviously not portable.

9.2 LARGE EQUIPMENT

Vehicle body repair workshops require many large pieces of equipment including spray booths, vehicle lifts, mixing rooms and body jigs. All of these require maintenance. Additionally, lifts, jacks and axle stands should be inspected for insurance

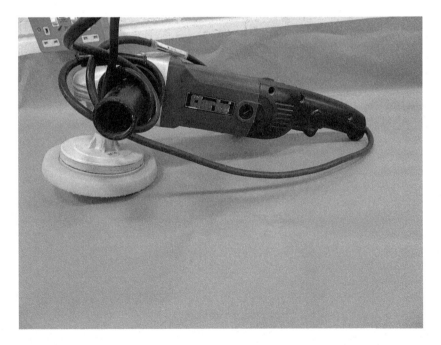

Figure 9.10 Electrical equipment should have the flex checked for damage before use. If it stops working, the first thing to check is the fuse. If the fuse is not blown check that the brushes are not excessively warn.

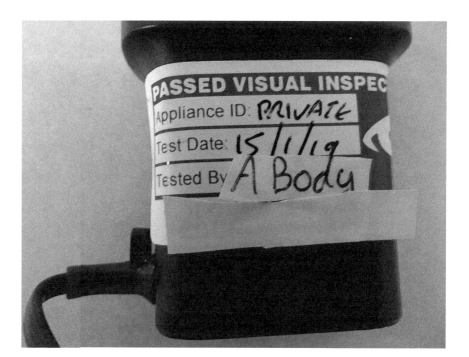

Figure 9.11 PAT labels let the people who use the equipment know when it requires retesting.

purposes, and any problems flagged up by an insurance inspector should be remedied to a suitable standard. Booth maintenance includes changing input and output filters, and cleaning the overspray from the walls.

These are both fairly simple procedures with a few stout tables as in Figure 9.12, especially cleaning the walls, but when changing input filters (pictured Figure 9.13), the operatives should have 'working at height' training.

The floor or wall extraction filters are simpler. Floor filters require the heavy metal grids featured in Figure 9.14 lifting carefully to gain access to the filters. It is important to understand the order in which extraction filters are layered. Figures 9.15, 9.16, 9.17 and 9.18 show the order in which the filters are layered up and how they are supplied on a roll. Floors with full extraction should have been balanced for consistent extraction across the entire filter surface. As can be seen in Figure 9.19 this appears as flat plates under the filters, with varying numbers of holes in them to regulate the extraction at every point. Lighting in the booth should be changed by someone with suitable training, as the electrical components have to be isolated from the booth atmosphere.

As well as changing the filters for a healthy atmosphere in the booth, it is also important to look after the compressed air supply. This is especially important for air-fed breathing equipment. Below are Figures 2.20, 2.21, 2.22 and 2.23 containing pictures of the transformer regulator which contains three filters in this example, but they are also available with just one or two filters. The three filters remove contaminants from the air supply more efficiently.

Figure 9.12 Here the spray booth has had the old sheeting removed and the walls washed down. Then fresh tacky coating was applied and a fresh stat sheet put up to protect it.

Figure 9.13 Input filters are found in the roof and require changing by people with working at heights training.

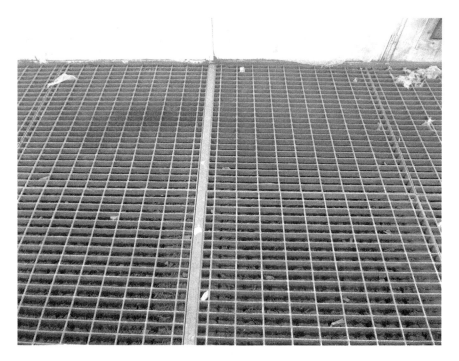

Figure 9.14 Extraction filters are found under the floor grates.

Figure 9.15 The course filters are placed green side up.

Figure 9.16 New floor filters are supplied on a roll. The green part is the coarsest filter.

Figure 9.17 Under the top filters are finer filters which further filter the air being extracted.

Figure 9.18 The finer filters are also supplied on a roll to be cut to the required length. They should be installed blue side up.

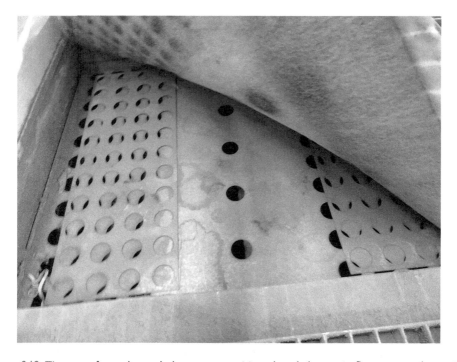

Figure 9.19 These perforated metal sheets are positioned to balance air flow across the entire booth floor.

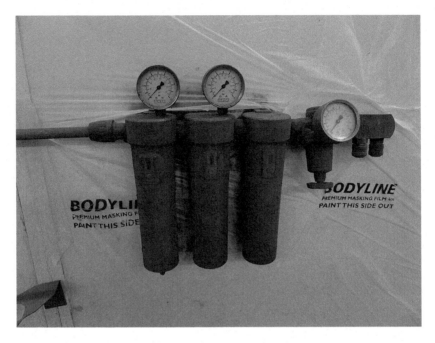

Figure 9.20 Transformer regulators may contain three filters which remove contaminants from the air supply.

Figure 9.21 Sintered filter removes oil and water.

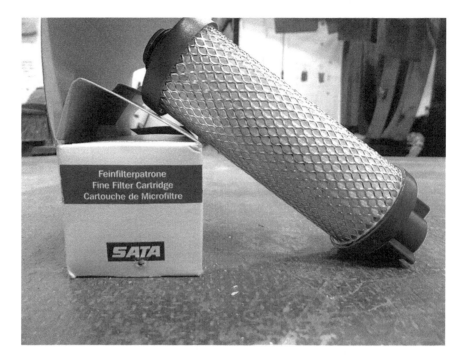

Figure 9.22 Fine filter removes particulates.

Figure 9.23 Activated charcoal filter removes oil vapour.

9.3 CLEANING APPLICATION EQUIPMENT

For efficient and correct application of all paints, spray guns should be kept in pristine condition. To do this effectively, it is important to understand how to strip the spray gun correctly, in order to access the components that require cleaning. For instance, the fluid tip should be removed after removing the needle and replaced before the needle. Please refer to the anatomy of a spray gun in Chapter 6 for full instructions.

It is important to identify what material has been through the spray gun. Two-pack materials should be cleaned out of the gun immediately with solvent gun wash. Solvent-based single-pack materials can be cleaned at a more relaxed pace, but again with solvent-based gun wash. Water-based paints should be cleaned out of spray guns with water-based gun wash. Many workshops opt to use tap water to clean spray guns, and this is acceptable, but the contaminated water should be collected and removed by a licenced waste disposal firm.

For other application equipment such as pressure-fed spray guns and suction-fed guns, please also refer to Chapter 6.

Chapter 10

Glossary

The refinishing industry is full of strange names and terms, and it is important to try and learn these in order to be able to converse with colleagues in your workshop or suppliers of paint materials. Local colloquialisms and slang terms will be avoided here as they could fill a book themselves. This chapter is broken down into subject areas:

10.1 TOOLS AND EQUIPMENT

- Aerosol – used for convenient delivery of foundation materials.
- Filler spreader – designed for the application of filler and stopper materials. Often given alternative uses like removing badges and trims, tucking masking into place and smoothing out stickers.
- Gravimetric – the process of mixing paint colours by accurately weighing separate tinters, using extremely sensitive scales – occasionally referred to as a 'gravimeter'.
- Heat lamp – a form of heating provided by infrared or near-infrared lamps. Used to direct heat at two-pack materials, in order to speed up curing.
- High volume low pressure (HVLP) – a term given to spray guns that deliver a high volume of paint material to the panel using a low air pressure. For a spray gun to be classed as HVLP, it must have a transfer efficiency of greater than 60%.
- Mixing board – a pad of nonabsorbent paper that is used to mix filler on, before the used sheet is torn off and thrown away.
- Mixing cup – used for mixing paint materials; these usually have mixing ratios and measurements printed on the side. Some are designed to be mounted straight onto the gun to replace the standard material cup.
- Mixing room – an extracted and ventilated room where paint is stored and mixed; usually contains the paint mixing scheme.
- Mixing scheme – the collective term for the storage shelves containing the tinters, the scales linked to a mixing computer and the colour chips or spectrophotometer.
- Mixing stick – used for stirring the paint. These may be aluminium with various ratios printed on them or plastic ones shaped in order to scrape tinters off the side of the mixing cup.
- Oven – the room in which vehicles or components are baked at around 70°C to speed up the cure time of two-pack materials.

- Panel stand – a piece of equipment of varying design, sometimes with hooks to hold panels at convenient heights and angles, often with wheels to help transport panels in and out of the spray booth.
- Safety data sheets (SDS) – used to give technicians, suppliers, couriers and disposal specialists safety information on the product.
- Spectrophotometer – an optical piece of equipment used to identify a vehicle's paint colour, via a link to a computer database of colours.
- Spray booth – the dedicated room where spraying should take place. Spray booths should have filtered extraction, to ensure that the air leaving the booth is as clean as possible. The air entering the booth should also be filtered, to provide a clean environment in which to paint. Most booths are also temperature controlled to provide an optimum environment for the application of paint. Combination booths also have a baking facility to speed up the curing of two-pack materials. This results in the booth often being referred to as the 'oven'.
- Spray gun – these include gravity-fed guns, where the material cup is above the gun; suction-fed guns, where the material cup is below the gun, and pressure-fed guns, where the material cup is held remotely from the gun, and the fluid is passed through a flexible hose to the gun, ready for atomisation.
- Technical data sheets (TDS) – the detailed instructions provided with paint materials.
- Transfer efficiency – the amount of paint that lands on the panel, having been atomised by the spray gun.
- Trestle – a padded tubular frame, which can be moved around the workshop and packed away when not in use. Used to support panels and components for prepping, filling, etc.

10.2 MATERIALS

- Abrasives – the collective name used to cover all materials designed to abrade or smooth down uneven surfaces, or remove gloss from painted surfaces.
- Adhesion promoter – a form of primer that is applied to bare plastic. This very thin solvent looks very similar to thinners and modifies the surface of the plastic to enable further foundation coats to or adhere to the substrate properly.
- Basecoat – the colour coats in clear-over-base systems. These may be metallic, pearl, mica, xirallic or straight colours. Basecoats may be solvent or water based; they provide cover over any primer repairs and are easily blendable. They do not provide gloss and are not waterproof.
- Clear coat – the waterproof transparent layer for clear-over-base systems. Usually glossy, but may be satin, or even matt. It provides a smooth, even coat over the entire panel, which provides a uniformity of blends.
- Direct gloss – any top coat that provides both colour and gloss at the same time.
- Elasticiser – also known as plasticiser or flexible additive. An additive usually for two-pack products such as primer or clear coat. Elasticisers enable paint layers to cure with a flexible property and are used on plastic panels.
- Etch primer – a foundation coat that provides both adhesion to bare metals and corrosion protection.

- Fade-out thinners – a form of thinners used to help the clear-coat edge flow out at the end of a clear-coat fade-out.
- Filler – a two-pack foundation material used to fill dents and to correct the contours on misshapen panels. It is sanded smooth to provide accurate shape and contours.
- Gun wash – a low-grade thinner used to clean solvent-based materials from application equipment. Should not be used for thinning paint products.
- High-build primer – a primer used to provide a degree of build to level slight undulations in panels. Like filler, it is sanded smooth after application.
- Lacquer – a name often used for clear coat. However, this is incorrect. Lacquers are single-pack products, usually cellulose or acrylic, which may possess colour or be clear.
- Masking paper – a special paper used to protect non-damaged parts of the vehicle from overspray. The paper is easy to cut and shape for different panels, and is available in a variety of sizes.
- Masking sheet – commonly referred to as 'stat sheet'. Masking sheet is a thin plastic cover that is provided folded up and on a roll.
- Masking tape – commonly available in various widths from 3 to 50 mm. The smaller sizes tend to be vinyl, while from 15 mm and up, they are paper. All have a low-tack glue on the reverse.
- Panel wipe – a solvent or water-based liquid used to remove grease, polish, skin oils and many other contaminants from panels.
- Polyester primer – a very high-build primer; technically speaking, a low viscosity filler, which is sprayed over the panel to provide a few millimetres of build.
- Primer surfacer – a medium-build primer, which still requires sanding like a high-build primer, but provides less build, more like a wet-on-wet.
- Stopper – a fine-grade, low-build filler for filling pinholes, sanding scratches and slight low spots in repairs.
- Tinters – the paints used to create the colour match for the vehicle. They have various metallic, pearl and coloured pigments.
- Volatile organic compound (VOC) – chemicals with a high-vapour pressure, detectable as a strong odour at normal temperatures.
- Wet-on-wet primer or non-sanding primer – a low viscosity primer applied over finely prepared panels in one smooth coat.

10.3 TECHNIQUES

- Back masking – this technique is widely used on the edges of panels. It is where the tape is stuck to the back of the edge of the panel and used to seal any overspray out from door shuts.
- Blending – the technique of fading colour out from full saturation over the repair to nothing before the end of the panel. This ensures that the panel should match the adjacent panel still after repair.
- Blow-ins – also known as fade-outs. This technique is used on gloss coats, either direct gloss or more usually clear coat, to enable the technician to not have to paint right up to the end of the panel, while still achieving a non-detectable repair.

- Colour matching – the process of using colour chips, spray-out cards or a spectrophotometer to obtain the correct colour to be used on a vehicle.
- Colour sanding – a term used to describe the final flatting of gloss coats after they have cured, in order to smooth out orange peel, dirt nibs and other defects in freshly applied paint, before it is finally polished to a desirable lustre.
- Edge-to-edge – the process of painting an entire panel (usually only used when discussing colour coats) from one edge to the other, rather than using blending techniques. The edge of the freshly painted panel will be close to the edge of an original panel.
- Fade-out – see blow-ins.
- Polish test – an experiment used to find out whether a paint finish is direct gloss or clear-over-base. Direct gloss will show colour on the polishing cloth, but clear-over-base will not change the colour of the cloth.
- SMART – a collective name for repairs where only a section of the panel is repaired. It stands for 'small medium area repair technique'.
- Solvent wipe test – a procedure used to determine whether a paint finish is single pack or two pack. Single-pack product will wrinkle up or smear when exposed to thinners. Two-pack product will be unaffected.

Index

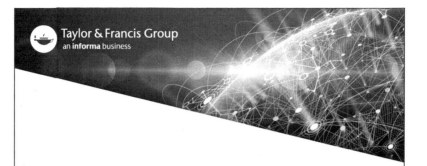